Total Internal Reflection Fluorescence (TIRF) and Evanescence Microscopies

About the Series

The Biophysical Society and IOP Publishing have forged a new publishing partnership in biophysics, bringing the world-leading expertise and domain knowledge of the Biophysical Society into the rapidly developing IOP ebooks program.

The program publishes textbooks, monographs, reviews, and handbooks covering all areas of biophysics research, applications, education, methods, computational tools, and techniques. Subjects of the collection will include: bioenergetics; bioengineering; biological fluorescence; biopolymers *in vivo*; cryo-electron microscopy; exocytosis and endocytosis; intrinsically disordered proteins; mechanobiology; membrane biophysics; membrane structure and assembly; molecular biophysics; motility and cytoskeleton; nanoscale biophysics; and permeation and transport.

A full list of titles published in this series can be found here: https://iopscience.iop.org/bookListInfo/iop-series-in-biophysical-society.

Total Internal Reflection Fluorescence (TIRF) and Evanescence Microscopies

Daniel Axelrod
Department of Physics, LSA Biophysics, University of Michigan, Ann Arbor, MI, USA

IOP Publishing, Bristol, UK

ISBN 978-0-7503-3351-1 (ebook)
ISBN 978-0-7503-3349-8 (print)
ISBN 978-0-7503-3352-8 (myPrint)
ISBN 978-0-7503-3350-4 (mobi)

DOI 10.1088/978-0-7503-3351-1

Version: 20220901

IOP ebooks

British Library Cataloguing-in-Publication Data: A catalogue record for this book is available from the British Library.

Published by IOP Publishing, wholly owned by The Institute of Physics, London

IOP Publishing, Temple Circus, Temple Way, Bristol, BS1 6HG, UK

US Office: IOP Publishing, Inc., 190 North Independence Mall West, Suite 601, Philadelphia, PA 19106, USA

Contents

Preface and acknowledgments

This book is intended as a 'one-stop shopping' exposition of the theory and setup details of evanescence microscopy—in particular, total internal reflection fluorescence (TIRF) and supercritical-angle fluorescence. It also includes discussions of many of the combinations of TIRF with other techniques. Every section 'starts from the beginning' so it should be readable by anyone with a science background that includes some physics and math. Most sections proceed to fairly advanced details. Hopefully, this will make the book usable as a research reference as well as a trigger for new ideas. In addition to the formal math, a physically intuitive explanation is provided wherever possible. Also, the topics and explanations here will hopefully not go out of date too soon, because they focus on the fundamentals underpinning the techniques. If any important techniques of optical evanescent microscopy have not been adequately covered, I apologize, and I would very much like to hear about them and possibly include them in future updates if necessary.

There is one thing that this TIRF book is *not*. It is *not* intended as a review of the many cell biology questions to which TIRF has been applied. The biological applications discussed here are included only because they illustrate the technique employed in studying them. Therefore, the reference lists ('Further Reading' lists at the end of each chapter) are limited to a subset of published works that provide more detail on the theory and techniques.

Many people have contributed to the writing in this book, whether intentionally or inadvertently. First, of course, are those people who are direct coauthors of particular sections, without whom those sections would not even exist here. This group (listed here with their most recent affiliations) includes my former graduate students in Physics at the University of Michigan, Nancy L Thompson (Dept. of Chemistry, University of North Carolina, Chapel Hill) and Edward H Hellen (Dept. of Physics, University of North Carolina, Greensboro). It also includes Geneva M Omann (Dept. of Biological Chemistry, University of Michigan) and Jeremy J Axelrod (Dept. of Physics, University of California, Berkeley). They were intentional contributors. An important but inadvertent contributor was Melvyn N Kronick, who I only fortuitously knew during a one week overlap while he was starting a postdoc at Berkeley just as I was finishing up there in Melvin P Klein's research group. Melvyn introduced me to the existence of TIRF (non-microscopic), the topic of his PhD thesis at Stanford in William Little's research group.

I also thank Dr Jerome Mertz (Boston University), Dr Paul Selvin (University of Illinois), Dr Bethe Scalettar (Lewis and Clark College), and Dr James Abney (Portland, OR) for useful discussions about diverse aspects of fluorescence microscopy, and Sharon Axelrod for help with a literature search of TIRF applications.

I have benefited from collaborations with many researchers at the University of Michigan, some of whom have been involved in TIRF applications with my group, and who generally provided the biological expertise that I lack. My longtime friend and colleague Ronald V Holz (Dept. of Pharmacology) was absolutely invaluable in enriching my scientific career (and extending it well past 'retirement'). Our postdocs

Miriam Allersma, Prabhodh Abbineni, Tania Chakrabarty, Annita Ngatchou-Weiss, Arun Anantharam, Vadym Degtyar, and Kevin Bohannon made important TIRF contributions.

Aside from the abovementioned former graduate students and partial coauthors Nancy Thompson and Ed Hellen, many of my (or shared) graduate students have made advances in TIRF microscopy, including Thomas P Burghardt, Robert M Fulbright, Ariane McKiernan, Andrea Stout, Michelle D Wang, Susan Sund Kulawik, Laura Nelson Johns, and Alexa Mattheyses. Other graduate students working in, or with, my group have developed other related aspects of fluorescence microscopy: Paul Selvin, Bethe Scalettar, Yifeng Yuan, Rhonda Dzakpasu, and Marisela Velez. Their contributions have increased my understanding of the field, and their presence has made it more fun.

The University of Michigan is a fantastic research center. Partly, this is because it is huge, such that it is hard to find a research field at the border of physics and biology in which somebody at UM is not an expert. But mostly, it is because long-term faculty and staff members there are so willing to collaborate and to be friends. Drs. Kate Barald, Richard Neubig, Miriam Greenberg, Joel Swanson, John Langmore, Jennifer Linderman, Mary Bittner, Bernhard Agranoff, and Edward Stuenkel, have all contributed their knowledge and/or lab facilities to the fluorescence microscopy work of my own group. I thank Dr Duncan Steel for helping, specifically with the multiphoton TIRF section. Special thanks go to Sharada Kumar for her years of invaluable help in keeping the lab at UM functioning smoothly. And I thank Dr Les Satin (of the U. Michigan Pharmacology Dept. and the Biophysical Society Publications Committee) for suggesting that I write this book.

Finally, several scientists from other institutions have provided helpful TIRF-related discussions over the years: Sanford Simon (Rockefeller University), Edwin Levitan (University of Pittsburgh), and especially Martin Oheim (University of Paris), who is a leading TIRF expert.

As can be inferred and extrapolated from the specific personal thanks in the above paragraphs, scientific progress depends on cooperation and collaboration. This is also true in a general sense, beyond the necessary (and enjoyable) personal interactions, because science is a public effort. Not only are some of the greatest universities in the world publicly owned and controlled (like U. Michigan), but the research itself is publicly funded almost everywhere. At the beginning of my scientific career at Berkeley, I wrote approximately the following in the acknowledgements section of my PhD thesis:

'Scientific exploration ideally is a very communal activity, where resources are commonly owned and shared and its fruits are openly disseminated, to be used for the better of the whole human community. Such an ideal structure would survive best in a society similarly structured. I thank progressive people everywhere for their active efforts to move reality and the ideal a little closer.'

Even umpteen years later, I still have the same sentiments.

Author biography

Daniel Axelrod

Daniel Axelrod grew up in Brooklyn, NY, USA and graduated from its public schools: PS 238, Brooklyn Technical High School, and Brooklyn College (B.S. in Physics and Math, 1968). He then went to graduate school at UC Berkeley (Physics Ph.D., 1974) where his thesis work (in the Laboratory of Chemical Biodynamics with Dr Mel Klein) involved the use of lanthanide fluorescence to study metal ion interactions during nerve action potentials. This was followed by an NIH Postdoctoral fellowship at Cornell University in Ithaca, NY, where he worked with Eliot Elson, Watt Webb, Dennis Koppel, Joseph Schlessinger, and Ken Jacobson to develop the fluorescence recovery after photobleaching (FRAP) microscopy technique. Finding an academic job back then was (and still is) difficult, so he applied and was accepted to medical school, just before a last-minute Assistant Professorship offer (in 1977) from the Physics Department and Biophysics Research Division at the University of Michigan in Ann Arbor. During his decades at UM, he received several awards for research (Biophysical Society Fellow, Biophysical Society Gregorio Weber award, for developing TIRF microscopy and related fluorescence microscopy techniques) and for teaching. Apart from straight physics and biophysics and microscopy courses, he also developed and taught multiple interdisciplinary full semester courses, including 'Physics of Music' and 'Science and Strategy in the Nuclear Arms Race'. As a Professor Emeritus starting in 2005, he has continued active optical biophysics research in collaboration with researchers at UM and elsewhere. In addition to academic work, Dan has also long been active in movements supporting environmental protection, peace and anti-militarism, and socioeconomic fairness.

Chapter 1

Introduction to optical evanescence

1.1 Overview

This book is intended as a guide to the physics, mathematics, and practical configurations of total internal reflection fluorescence (TIRF) microscopy and related microscopies that depend on optical evanescence and also their combinations with other microscopic techniques. As such, the theory is emphasized, with its rigorous math. However, at every opportunity, a physically intuitive interpretation is explored. Also emphasized are the various ways in which TIRF and related techniques are implemented. The shortcomings and limitations in the range of applicability of TIRF, such as scattering, are also discussed in mathematical detail. Future possible directions of TIRF and other evanescence techniques are also mentioned.

The mathematics assumes a familiarity with vector calculus and partial differential equations. Otherwise, the presentation is intended to be self-contained and complete and therefore recapitulates some material found in standard electricity and magnetism textbooks.

The intensity of evanescent light decays rapidly away (typically in less than one wavelength) in at least one spatial dimension. TIRF depends on engineering the optical path to produce evanescence of the light that excites a fluorescent sample. Evanescence occurring in fluorescence emission itself (as opposed to excitation), such as in the near field of fluorescing molecules, is closely related to the principles of TIRF. Evanescence can also be produced and implemented in other optical systems, such as near-field scanning microscopy (NSOM) and waveguides. Those systems are covered here, but in less detail.

In all cases, the purpose of using evanescence is to concentrate the light on the portion of the sample of particular interest, without exciting background fluorescence (usually out-of-focus) from other regions. This greatly enhances image contrast.

doi:10.1088/978-0-7503-3351-1ch1

1.2 Applications to biochemistry and cell biology

Some of the general areas in biochemistry and cell biology in which TIRF microscopy has been found useful are as follows:

1.2.1 Cell/substrate contact regions

TIRF can be used qualitatively to observe the position, extent, composition, and motion of contact regions, even in samples in which fluorescence elsewhere would otherwise obscure the fluorescent pattern. The specific biochemistry of focal adhesions has been studied using TIRF.

A variation of TIRF used to identify cell–substrate contacts involves doping the solution surrounding the cells with a nonadsorbing and impermeable fluorescent volume marker. Focal contacts then appear relatively dark.

1.2.2 Long-term videos of living cells

Epi-fluorescence illumination appears to adversely affect cell viability, even with intermittent exposure for making time-lapse movies. TIRF seems particularly advantageous for the long-term viewing of cells, since the evanescent wave minimizes exposure of the cells' organelles to excitation light.

1.2.3 Secretory granule tracking and exocytosis

Secretory granules often reside through the whole depth of a cell. When viewed using TIRF, the granules closest to the membrane appear very distinctly (figure 1.1, top row). When viewed with standard epi-illumination (EPI), fluorescence-marked secretory granules are difficult to distinguish individually (figure 1.1 bottom row).

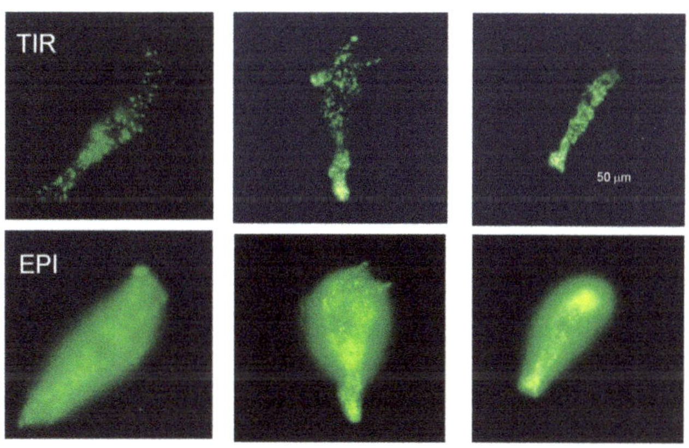

Figure 1.1. Objective-based TIRF vs EPI with NA = 1.65 on GFP-marked secretory granules in bovine chromaffin cells. Reprinted with permission from Axelrod (2001), CC-BY 4.0.

The thin evanescent field allows small motions of individual fluorescence-marked secretory granules in the direction normal to the substrate and plasma membrane (the 'axial' or 'z' direction) to manifest as small intensity changes. The precision of such axial tracking can be as small as 2 nm, which is much smaller than the light microscope resolution limit. Because of the high contrast and low background, the positions of the centers of granules can be measured with an accuracy of down to about 10 nm, and motions much smaller than the granule diameter can be followed before and after exocytosis-inducing chemical stimulation. In some cases, the dispersal of granule contents can be observed and interpreted as exocytosis, and events immediately preceding and following exocytosis can be characterized.

Most of the TIRF studies of the detailed mechanism of secretion have used animal cells, but the same techniques can be applied to plant cells.

1.2.4 Single molecules

The process of imaging single molecules has always been somewhat challenging because of the finite number of photons (generally in the many thousands) that can be gathered from an organic fluorophore before it irreversibly photobleaches. However, the main challenge has been to reduce out-of-focus background, because light from any source, even extraneous sources, produces shot noise. TIRF has therefore become a mainstay of single-molecule imaging.

Even with TIRF background suppression, the ability to 'see' a single molecule (really, to see the center of its location, not its details) is a question of creating a distinct, diffraction-limited spot that does not overlap with its neighbors. As a rough approximation, a single molecule can be distinguished if it is the only fluorescent occupant within a resolution-limited volume. For TIRF, that volume is very small: it is roughly the product of the evanescent field depth (~80 nm) multiplied by a lateral area that has a radius of ~200 nm (an Airy disk point spread function), i.e. a volume of $\sim 10^7$ nm^3. For fluorophores in a uniform solution (near the total internal reflection (TIR) surface), this corresponds to a maximum bulk concentration of ~100 molecules/μm^3. Indeed, when setting up TIR, the transition from a subcritical to a supercritical illumination angle produces a dramatic transition from smoothly diffuse fluorescence from a fluorescent solution to individual single-molecule discrete spots, twinkling as single molecules diffuse in and out of the evanescent field.

A substantial out-of-focus background (without TIRF) can arise from *in vitro* samples where the binding of a protein of interest to a surface-bound target is reversible. In this case, much of the protein is present in equilibrium in an unbound state deeper in the solution. However, with TIRF, binding and unbinding rates can be quantified either by fluorescence correlation spectroscopy (TIR/FCS) or by direct temporal viewing of the appearance and disappearance of single molecules as they park or wander through the evanescent field or diffuse away to invisibility in the solution.

Out-of-focus background can be even more severe *in vivo*, in intact cells, because some intrinsically fluorescent biochemicals have a 'tail' in their excitation and emission spectra that overlaps with the spectra of the fluorescence-labeled protein of interest. In addition, much of the labeled protein of interest (either introduced exogenously through the plasma membrane or by genetic transfection) may reside at locations in the cytoplasm that are not of interest for a specific target. Since TIRF largely overcomes this background problem for intact cells, it has been used to study, for example, single-ion channels and G-proteins in plasma membrane and virus particles as they infect cells.

1.2.5 Reversibly bound and mobile fluorescent ligands on cells and biosurfaces

Many cell–surface receptors bind their ligands reversibly, so a surface-bound population can be maintained only with a substantial concentration in the bulk. In some other cases, the cell-surface receptor binds ligands irreversibly, but then the complex is internalized and replaced by a fresh unbound receptor. The newly appearing receptors can be labeled continuously by fluorescent ligand in the bulk. The bulk fluorescence is discriminated against by TIRF.

If the evanescent wave intensity is briefly brightly flashed, then some of the fluorophores associated with the surface become photobleached. Subsequent exchange with unbleached dissolved fluorophores in equilibrium with the surface leads to a recovery of fluorescence, excited by a continuous but much attenuated evanescent wave. The time course of this recovery is a measure of the desorption kinetic rate. This technique is called TIR/FRAP in reference to 'fluorescence recovery after photobleaching.' Kinetic rates of hormones and cytoskeletal proteins at model membranes, flattened biological membranes, and intact cell surfaces have been measured by this technique. Kinetics at the cytofacial surface of intact cells can be viewed by employing TIR/FRAP on cells that have been microinjected with a labeled molecule of interest.

TIR/FRAP also can be used to measure surface diffusion coefficients along with on/off kinetics, if the evanescent wave intensity varies over a short distance on the surface compared to the characteristic distance covered by surface diffusion within the time available before desorption. This variation in TIR illumination intensity can be introduced by focusing, by interfering two TIRF beams, or by placing a striped pattern ruling in the incident beam at a position where it forms a real image on the sample.

TIR/FRAP can detect lateral movement of fluorescent-marked transfected proteins on (or in) submicroscopic-sized organelles, such as secretory granules. Secretory granules are about 300 nm in diameter, so TIR (with an evanescent field depth of 50–100 nm) can be used to illuminate (and photobleach) just one portion of a granule, i.e. the part closest to the TIR substrate. Postbleach redistribution of the fluorophores then provides a measure of the diffusion coefficient of the labeled protein on the granule membrane or within its interior.

Also related to TIR/FRAP is the capability of seeing fluorescence fluctuations as fluorescent molecules enter and leave the thin evanescent field region in the bulk. These visually obvious fluctuations can be quantitatively autocorrelated to obtain kinetic information about the molecular motion (fluorescence correlation spectroscopy (FCS) and in this case, TIR/FCS)

1.2.6 Cytoplasmic filaments

Although TIRF cannot obtain a deep view of thick cells, it can offer a high-contrast view of the fluorescence-marked submembrane filament structure at the substrate contact regions. Actin polymerization and its regulation, both *in vitro* and *in vivo*, have been studied by TIRF, as have microtubules and microtubule-associated kinesin. The manner by which extracellular components such as fibronectin are organized through the membrane in concert with submembrane filaments can be investigated by TIRF. In combination with very low levels of specific filament labeling, which produces fluorescent speckles, TIRF illumination can visualize filament subunit turnover.

1.2.7 Calcium channels and transients

TIRF is completely compatible with standard epi-fluorescence, bright-field, dark-field, or phase-contrast illumination. These methods of illumination can be switched back and forth rapidly with TIRF by electro-optic devices to compare membrane-proximal ionic transients vs. deeper cytoplasmic transients, which is particularly useful for following calcium waves.

1.2.8 CRISPR

TIRF microscopy has become popular in recent years, in part because its ability to greatly reduce unwanted fluorescence background has coincided well with the new development of highly specific labeling techniques utilizing genetic manipulation: e.g., green fluorescent proteins (GFPs) and Clustered Regularly Interspaced Short Palindromic Repeats (CRISPR). This coincidence has opened new possibilities for single-molecule and organelle tracking in cell biology. Fluorescent labeling of specific proteins in living cells, which are then tracked by TIRF, has become quite versatile when used with modern CRISPR techniques (Matsuda and Oinuma 2019).

The interactions of specific proteins generated by the CRISPR technique with DNA molecules covalently attached to the surface of a microfluidic chip can be examined by prism-based TIRF and appropriate image analysis software in a massively parallel manner (Jung *et al* 2017).

1.2.9 Orientational distributions of fluorescent molecules at a surface

The polarization properties of the evanescent wave can be used to excite selected orientations of fluorophores. Standard polarized epi-illumination cannot distinguish order from disorder in the polar direction (defined as the optical axis), because epi-illumination is polarized transverse to the optical axis. However, microscope TIR illumination can be partially polarized in the optical axis direction (the z-direction) and can thereby detect order in the polar direction. If a fluorophore is incorporated into a lipid membrane in a restricted orientation, the fluorophore ordering detected by polarized TIRF can result directly from folding and indentations in the membrane, even at a submicroscopic level.

1.2.10 Combinations and comparisons with other microscopy techniques

TIR has been combined with fluorescence lifetime imaging of cultured cells. The labels range from terbium chelates, with very long (millisecond) lifetimes, to organic fluorophores with nanosecond lifetimes.

TIRF is optically compatible with the much higher-resolution scanning force microscopy (SFM) technique, in which mutual registration of the two types of image requires that both types be acquired on the exact same sample of surface-adsorbed single molecules. TIRF is also compatible with the photo-activated localization microscopy (PALM) and stochastic optical reconstruction microscopy (STORM) super-resolution techniques, which feature outstanding lateral resolution at the expense of speed and the use of TIRF to augment the axial resolution.

Confocal microscopy (CM) is another technique for apparent optical sectioning, which is achieved by excluding out-of-focus emitted light using a set of image-plane pinholes. CM has the clear advantage in versatility; its method of optical sectioning works at any plane of the sample, not just at an interface between dissimilar refractive indices. However, other differences exist which, in some special applications, can favor the use of TIRF:

- The depth of the optical section in TIRF is ~0.1 μm, whereas in CM it is a relatively thick ~0.6 μm.
- In some applications (e.g. FRAP, FCS, or on cells whose viability or fluorescence is damaged or decremented by light), illumination and not just detected emission is best restricted to a thin section; this is possible only with TIRF.
- Since TIRF can be adapted to and made interchangeable with existing standard microscope optics, even with homemade components, it is much less expensive than CM.

Cell–substrate contacts can be located by a nonfluorescence technique completely distinct from TIRF, known as 'internal reflection microscopy' (IRM). Using conventional illumination sources, IRM visualizes cell–substrate contacts as dark regions. IRM has the advantage that it does not require the cells to be labeled, but it has the disadvantages that it yields no information about biochemical specificities in the contact regions and that it is less sensitive to changes in contact distance (relative to TIRF) within the critical first 100 nm of the surface.

Structured illumination microscopy (SIM), which illuminates the sample with, for example, stripes, provides optical sectioning of a fluorescent sample in the axial direction (normal to the sample coverslip surface) and additionally improves lateral resolution by about a factor of two. Its disadvantage is that it is essentially a post-acquisition technique: at least nine images, which must be of the same field, with differently oriented stripes and phase-shifted stripes, are then subjected to a mathematical treatment. However, SIM can be combined with TIR to produce resolutions in all three dimensions that are considerably enhanced compared to those of either standard SIM or TIRF alone. TIR-SIM is discussed in more detail in section 5.11.

Supercritical angle fluorescence (SAF) microscopy also provides optical sectioning. Unlike TIRF, SAF is based on the evanescent properties of fluorescence emission rather than excitation illumination. However, both SAF and TIRF have overlapping physical theories of evanescence, and their combination produces even higher resolution than either one alone. This subject is covered in section 5.8.

1.3 Ray picture of total internal reflection

Most of the theory presented in this book necessarily considers electromagnetic waves. However, the simplest view of TIR (figure 1.2) only needs to consider 'rays,' which represent the direction that light travels normal to its wave fronts.

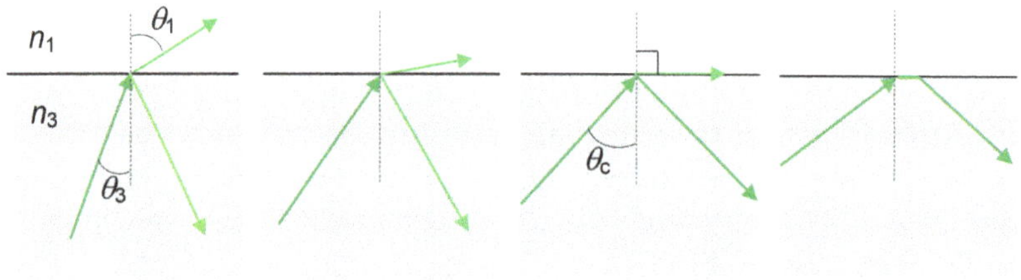

Figure 1.2. Ray diagram for refraction at an interface between media with refractive indices n_3 and n_1, where $n_3 > n_1$. The incidence angle increases from left to right.

The incidence angle θ_3 and the refracted angle θ_1 are related by Snell's law: $n_1 \sin\theta_1 = n_3 \sin\theta_3$. At the 'critical' incidence angle θ_c, the refracted ray skims for an 'infinite' distance along the surface at $\theta_1 = 90°$. As θ_1 increases beyond θ_c, the refracted ray skims the surface for a distance (called the Goos–Hänchen shift) that rapidly decreases with increasing supercritical incidence before 100% of it is reflected back into n_3.

1.4 Maxwell's equations and wave numbers

Rays adequately describe the existence of TIR, which occurs if the incidence angle exceeds the critical angle, but they say nothing about the evanescent field in the lower-density medium. The mathematical and physically intuitive groundwork for evanescence can described by starting from the beginning: Maxwell's equations.

Maxwell's equations for a linear medium with no free charges or currents are:

$$\begin{aligned} &\nabla \cdot (\varepsilon \mathbf{E}) = \mathbf{0} \\ &\nabla \cdot \mathbf{B} = \mathbf{0} \\ &\nabla \times \mathbf{E} = -\frac{\partial \mathbf{B}}{\partial t} \\ &\nabla \times \frac{\mathbf{B}}{\mu} = \frac{\partial (\varepsilon \mathbf{E})}{\partial t}, \end{aligned} \qquad (1.1\text{ a–d})$$

where ε is the dielectric permittivity. In the section on scattering (section 2.4), we will allow the dielectric constant to vary over space, but here we will assume $\varepsilon = \varepsilon_1$ is constant over space in a region '1,' but possibly different than it would be in a vacuum, where $\varepsilon = \varepsilon_0$. We will assume the medium is nonmagnetic, so the magnetic permittivity $\mu = \mu_0$ (constant over all space). To eliminate **B**, we take the time derivative of equation (1.1d)

$$\left(\nabla \times \frac{\partial \mathbf{B}}{\partial t}\right) = \varepsilon_1 \mu_0 \frac{\partial^2 \mathbf{E}}{\partial t^2} \tag{1.2}$$

and then use equation (1.1c):

$$-(\nabla \times \nabla \times \mathbf{E}) = \varepsilon_1 \mu_0 \frac{\partial^2 \mathbf{E}}{\partial t^2}. \tag{1.3}$$

We now use the vector calculus identity $\nabla \times \nabla \times \mathbf{E} = \nabla(\nabla \cdot \mathbf{E}) - \nabla^2 \mathbf{E}$ along with equation (1.1a) to get:

$$\nabla^2 \mathbf{E} - \varepsilon_1 \mu_0 \frac{\partial^2 \mathbf{E}}{\partial t^2} = 0. \tag{1.4}$$

This is the standard wave equation

$$\nabla^2 \mathbf{E} - \frac{1}{v^2} \frac{\partial^2 \mathbf{E}}{\partial t^2} = 0, \tag{1.5}$$

where the speed of the wave v is

$$v = \frac{1}{\sqrt{\varepsilon_1 \mu_0}} = \frac{1}{\sqrt{\varepsilon_0 \mu_0}\sqrt{\dfrac{\varepsilon_1}{\varepsilon_0}}} = \frac{c}{n_1}, \tag{1.6}$$

and where $c = (\varepsilon_0 \mu_0)^{-1/2}$ is the speed of light in vacuum, and $n_1 \equiv (\varepsilon_1/\varepsilon_0)^{1/2}$ is the definition of the refractive index of region 1 ('medium 1').

Equation 1.5 is satisfied by *any* wave shape, as long as it preserves its shape as it propagates. However, we will choose a particular subset of shapes—single frequency planar sine waves—because: (a) any other shape can be constructed by a superposition of planar sine waves; (b) they are easily manufactured in the physical world by radiation from lasers and sinusoidally oscillating dipoles; and (c) because each frequency's sine-wave solution corresponds to a particular color and most of the presentation here will refer to a single color (i.e. monochromaticity). Planar sine waves have the mathematical form

$$\mathbf{E} = E e^{i(\mathbf{k}_1 \cdot \mathbf{r} - \omega t)} = E e^{i\left(k_x x + k_y y + k_z z\right)} e^{-i\omega t}, \tag{1.7}$$

where $\mathbf{k}_1$ is the 'wave vector' in medium 1 that points in the direction of propagation of the plane wave (with rectilinear components k_x, k_y, k_z, with subscript '1' on each implied but suppressed for conciseness) and with an amplitude that depends on the sine-wave frequency:

$$|\mathbf{k}_1| \equiv k_1 = \left(k_1^2\right)^{1/2} = \left(k_x^2 + k_y^2 + k_z^2\right)^{1/2} = \frac{2\pi}{\lambda_1} = \frac{n_1 \omega}{c}, \tag{1.8}$$

where λ_1 is the wavelength (the spacing between successive peak-to-peak wave fronts) for propagating light in medium 1, and ω is the frequency (in radians/sec) of the light. The key point here is that, in any isotropic and homogeneous medium at any particular frequency, ${k_1}^2$ must be a fixed real positive scalar in order for the planar sine wave to be a solution of the wave equation.

1.5 Causes of evanescence: a physical view

For 'normal' propagating planar sine waves, each of (k_x, k_y, k_z) is real and *each is less than* k_1. The sum of their squares is k_1^2, in accordance with the Pythagorean theorem in real 3D space. Figure 1.3 graphically shows this relationship for the pictorially simplified case of plane wave wave fronts restricted to propagation in the x–z plane (so $k_y = 0$). As the angle of propagation of $\mathbf{k}_1$ changes in this 2D example, both k_x and k_z change *oppositely*: when one increases, the other decreases.

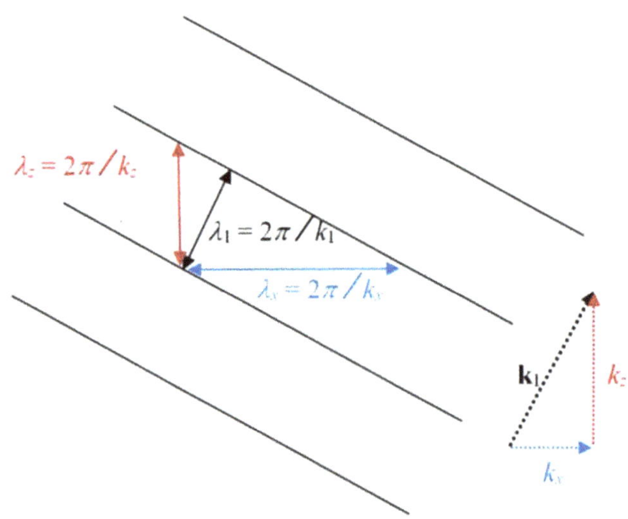

Figure 1.3. Wave fronts of a plane wave propagating to the upper right in medium 1. 'Wave front' is defined as the points of equal phase in a sine-wave variation.

However, there is nothing in the theory of Maxwell's equations that prevents any one (or two) of the three components of $\mathbf{k}_1$ from being made larger than k_1. All that is necessary to physically accomplish that is to somehow force the peak-to peak distance between successive wave fronts to be less than λ_1. Assume for simplicity that this squeezing occurs in the x–y plane. Then $k_x^2 + k_y^2 > k_1^2$. But according to equation (1.8), we then must have $k_z^2 < 0$. Thus k_z is forced to become completely imaginary, so that equation (1.7) becomes

$$\mathbf{E} = Ee^{i(k_x x + k_y y)}e^{-i\omega t}e^{-|k_z|z}. \quad (1.9)$$

From a physically intuitive point of view, 'squeezing' the wave fronts in the x or y directions beyond the minimum allowable in the medium 1, which is λ_1, then *forces* the z-direction dependence to become an exponential decay. 'Evanescent' light, in this book, means light that is forced into an exponential decay in a least one dimension by this mechanism, 'propagating' light refers to light that is not evanescent.

The imaginary nature of k_z transforms the Pythagorean relationship to $k_x^2 = k_1^2 + |k_z|^2$. So in this evanescent case, both k_x and $|k_z|$ change *similarly*: when one increases, the other also increases. The purely propagating planar wave depiction in the above figure transforms to figure 1.4:

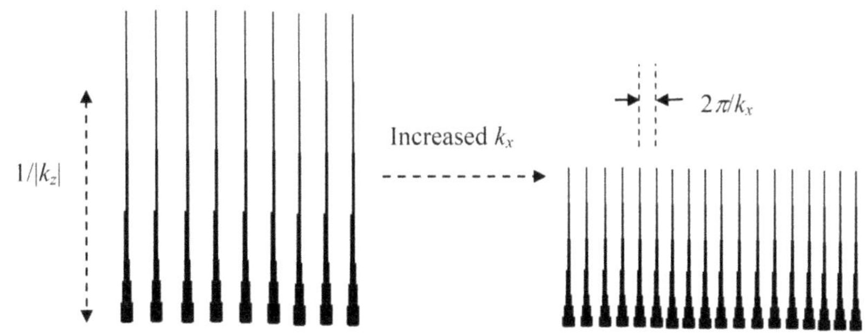

Figure 1.4. Schematic view of the relationship between k_x and $|k_z|$ for evanescent light.

In every case in optics in which an exponential decay is encountered in at least one dimension of a light field, the wave fronts of at least one of the other dimensions are being squeezed together so that they are closer than λ_1. As we will see, this general principle accounts for evanescence in each of the situations discussed in detail below: (a) total internal reflection; (b) small apertures; (c) waveguides, and (d) the vicinity of molecular-size fluorescence emitters.

1.5.1 Total internal reflection

The first example of wavelength 'squeezing' to produce evanescence is in TIR, depicted schematically in figure 1.5

Figures 1.5(a)–(c) show the progression of wave fronts as the angle of incidence θ_3 is increased and passes through the critical angle. In the subcritical range, increasing θ_3, forces the wave fronts in medium 1 to increasingly be oriented normal to the interface, while still retaining their spacing λ_1. Further increasing θ_3 forces the wave fronts in medium 1 to be closer together than their propagation wavelength λ_1, because this is the only way that the wave fronts can maintain the same spacing λ_x as that in medium 3, as measured along the interface. The closely-spaced wavefronts do move along the interface, in this case along x from left to right, at a speed such that the frequency is still unaltered from the incident light in medium 3. In this book, that evanescent wavefront motion is called 'travel' rather than 'propagation', to distinguish it from non-evanescent light.

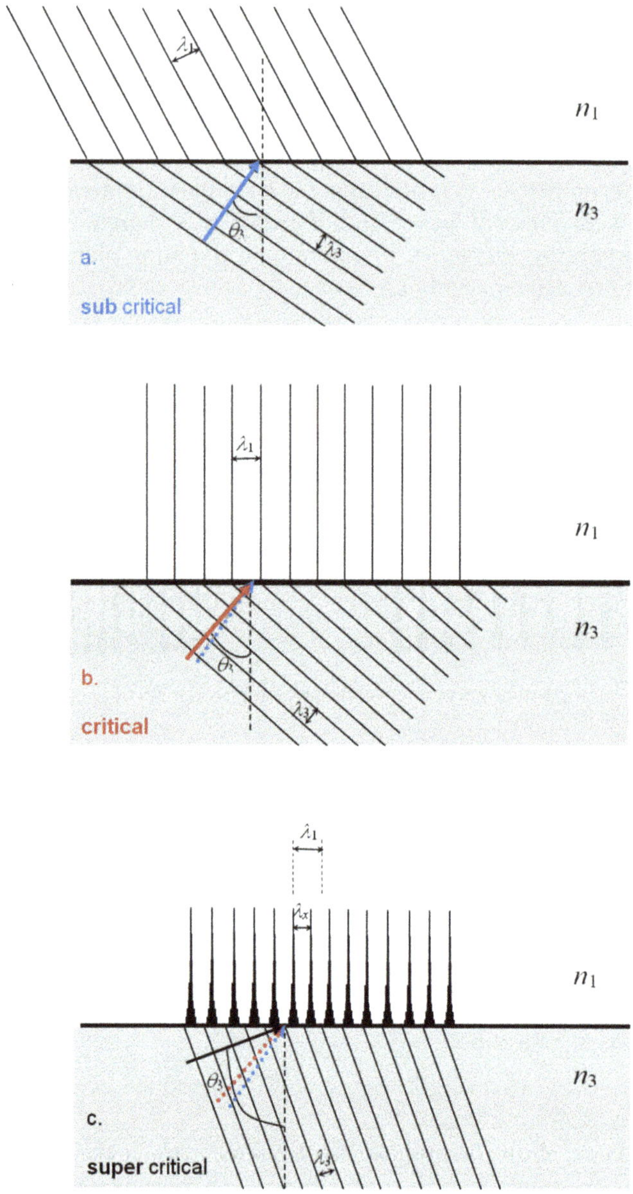

Figure 1.5. Schematic views of wave fronts at an $n_1 < n_3$ interface for (a) subcritical (blue arrow), (b) critical (red arrow), and (c) supercritical (black arrow) angles of incidence θ_3. Supercritical incidence produces an evanescent field in medium 1.

1.5.2 Small aperture

'Squeezing' of the wave front spacing to produce evanescence can also be accomplished by other means. Because there is no interface involved, we will drop the subscript '1' on the wavenumbers k. In the case of light passing through a small aperture embedded in a medium 1, λ_x squeezing can be understood by a Fourier component analysis. Consider a 1D aperture (a slit) for simplicity, with a plane wave incident normally upon it (i.e. at an incidence angle of zero). This sets up a standard 'single-slit' diffraction pattern many wavelengths away, but here we are concerned with the range of λ_x values immediately beyond the slit. Just before the slit we have $\lambda_z = \lambda_1$ and $\lambda_x = \infty$.

The wave front just beyond the slit has an electric-field amplitude $E(x)$ (normalized) which can be expressed in terms of its Fourier transform:

$$E(x) = \begin{cases} 1 & -a < x < a \\ 0 & \text{elsewhere} \end{cases}$$

$$= \frac{1}{\sqrt{2\pi}} \int_{-\infty}^{\infty} \tilde{E}(k_x) e^{ik_x x} dk_x = \frac{2}{\sqrt{2\pi}} \int_0^{\infty} \tilde{E}(k_x) \cos(k_x x) dk_x$$

where (1.10)

$$\tilde{E}(k_x) = \frac{1}{\sqrt{2\pi}} \int_{-\infty}^{\infty} E(x) e^{-ik_x x} dx = \frac{1}{\sqrt{2\pi}} \int_{-a}^{a} e^{-ik_x x} dx = \frac{1}{ik_x\sqrt{2\pi}} \left(e^{ik_x a} - e^{-ik_x a}\right)$$

$$= \frac{2}{\sqrt{2\pi}} \left(\frac{\sin k_x a}{k_x}\right)$$

The finite-width rectangular profile of $E(x)$ can only be constructed using a wide range of spatial frequencies k_x of cosine waves. Some of these frequencies, namely the low ones in the range $0<k_x<k_1$ (where $k_1 = 2\pi/\lambda_1$ is the spatial frequency of the incident light), produce real k_z frequencies in the range $0<k_z<k_1$ according to the Pythagorean $k_z^2 = k_1^2 - k_x^2$, representing propagating light traveling away from the slit at some angle α from the slit (where $\alpha = \tan^{-1}(k_z / k_x)$). But the higher k_x frequencies, i.e. those for which $k_x>k_1$, necessarily produce imaginary k_z values and thereby evanescence.

The border between propagating and evanescence contributions to the field beyond the slit lies at $k_x = k_1$, regardless of the aperture size a. However, the amplitude of those contributions strongly depends on a through the $\sin(k_x a)/k_x$ factor. If a is very large compared to λ_1, this factor is small for the evanescent contributions ($k_x>k_1$), and virtually all of the light is propagating. Conversely, for

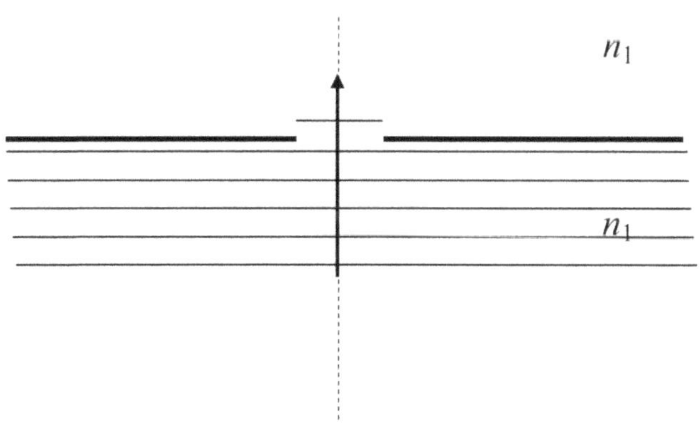

Figure 1.6. Light wave fronts passing through a small aperture.

$a << \lambda_1$, most of the light emerging from the slit is evanescent and decays rapidly with the distance from the slit. The dim propagating contribution, cast upon a distant screen, is the classic single-slit diffraction pattern.

The decay with z immediately after the single slit is not a single exponential but rather a continuous mix of characteristic decay $|k_z|$ values weighted by the $\sin(k_x a)/k_x$ factor. Another variable that affects the decay is the angle of incidence at the aperture. It need not be zero, as assumed here for simplicity.

1.5.3 Waveguides

A third type of k_x squeezing, related to the one above, occurs in a waveguide. This is relevant to microscopy because 'near-field scanning optical microscopy' uses a thin optical fiber with a very small aperture for raster scanning very near the sample. The optical fiber is essentially an aperture at the end of a waveguide, so the above considerations are qualitatively relevant. However, an optical fiber only supports certain modes (combinations of wavelengths and angles) for light traveling along its axis, so the light reaching the aperture sets up different weightings of evanescent decay lengths. Only some incidence angles are permitted.

1.5.4 Near-field emission

Finally, k_x squeezing occurs in the emission field of a fluorophore and leads to what is called the 'near field.' This field is describable by a classical approximation, in which the emission pattern is seen to be caused by, and follows the profile of, a tiny oscillating electric dipole of a size much smaller than the wavelength of the propagating light (the 'far field') it produces.

Chapter 4 discusses in detail the quantitative mathematics of this emission field. The field at the position of the dipole ($r = 0$) is a theoretical 'singularity'; it approaches infinite amplitude. At distances r much larger than the propagating emission wavelength, the field amplitude drops off with a $1/r$ dependence. The energy flux (watts/area) away from the dipole is given by the Poynting vector, which is proportional to the local $|E|^2$, so the 'far field' flux drops off with the familiar $1/r^2$ dependence. If we draw a hypothetical large sphere around the dipole, the total outward energy flow through that sphere is constant, regardless of radius. The 'near field' provides additional electric-field amplitude near the dipole, but does not contribute to the long-distance outward energy flow.

In chapter 4, we will discuss how an electromagnetic field can be expressed as an 'angular spectrum of plane waves.' The total field around a dipole cannot be written as the sum (or integral) of superimposed plane waves that all have the same wavelength spacing, regardless of their 'angular spectrum' of propagation directions. To account for the singularity at $r = 0$, we must include plane waves of shorter spacing, including some that lead to evanescence. This evanescence creates interesting optical effects when a fluorophore is near enough to an interface that can convert the evanescent fields to propagating light.

Further reading

Axelrod D 2001 Selective imaging of surface fluorescence with very high aperture microscope objectives *J. Biomed. Opt.* **6** 6–13

Axelrod D 2022 Total internal reflection fluorescence microscopy *Encyclopedia of Cell Biology* 2nd edn ed R A Bradshaw, G W Hart and P D Stahl (Amsterdam: Elsevier-Academic Press)

Jung C *et al* 2017 Massively parallel biophysical analysis of CRISPR/Cas complexes on next generation sequencing chips *Cell* **170** 35–47

Matsuda T and Oinuma I 2019 Imaging endogenous synaptic proteins in primary neurons at single-cell resolution using CRISPR/Cas9 *Mol. Biol. Cell* **30** 2838–55

Chapter 2

Total internal reflection theory

Before discussing the role of evanescent fields in exciting fluorescence emission, or the near-field evanescence of the emission itself, we first consider in more theoretical detail the creation of evanescence by total internal reflection at an interface between media of different refractive indices. Even without fluorescence, TIR has interesting, important, and practical features. The emphasis here is on the behavior of plane waves at the interface, for which the direction of propagation is characterized as a 'ray' or 'light beam.' In this chapter, we consider only a 'simple' interface between two semi-infinite but otherwise homogeneous media. In chapter 3, we consider the more complicated (but entirely common) effects of spatial variations and/or light absorption on the refractive index of the lower-index medium.

2.1 Rays and TIR

When a light beam (see figure 1.2) propagating through a transparent medium (known here as 'medium 3') that has a high refractive index (e.g. glass) encounters a planar interface with a medium ('medium 1') that has a lower refractive index (e.g. water), it undergoes transmission (also known as refraction) and reflection according to Snell's Law:

$$n_1 \sin \theta_1 = n_3 \sin \theta_3, \tag{2.1}$$

where n_1 and n_3 are the refractive indices of the liquid and the solid, respectively, and the angles θ_1 and θ_3 are measured from the normal to the interface, as in figure 1.5. ('Medium 2' and index n_2 are kept in reserve for intermediate layers, as discussed in section 3.2.) Snell's Law is a straightforward geometrical consequence of the boundary condition requirement that the wavefronts of the incident and transmitted plane-wave fields have the same wavefront spacing in the direction along the interface in the plane of incidence (here, x).

doi:10.1088/978-0-7503-3351-1ch2

If $n_1 < n_3$, an angle $\theta_3 = \theta_c$ exists such that $\theta_1 = \pi/2$, i.e. the transmitted beam just skims along the surface. This is the same situation as shown in figure 1.5(b), where the wavefronts are perpendicular to the surface. That 'critical angle' θ_c is given by:

$$\theta_c = \sin^{-1}(n_1/n_3). \tag{2.2}$$

At angles $\theta_3 > \theta_c$, the incident beam is totally reflected and the transmitted beam is evanescent, as shown in figure 1.5(c).

What happens to θ_1, the angle of refraction, once we go beyond the critical angle, $\theta_3 = \theta_c$? From Snell's Law, we can derive a general expression for $\cos\theta_1$:

$$\cos\theta_1 = \sqrt{1 - \sin^2\theta_1} = \sqrt{1 - (n_3/n_1)^2 \sin^2\theta_3}. \tag{2.3}$$

For $n_3 > n_1$ and $\theta_3 > \theta_c$ (the TIR criteria), $\cos\theta_1$ becomes (positive) imaginary. This feature is central to the exponential decay, amplitudes, and the special phase and polarization features of TIR evanescence.

2.2 Waves and TIR

The above elementary ray analysis says nothing about the amplitudes or phases of the transmitted (i.e. refracted) and reflected beams. For a more complete picture, we use Fresnel coefficients, which are derived from boundary conditions imposed on the solutions of Maxwell's equation for the interface problem.

We assume here that the light is confined to the x–z plane (the plane of incidence), so that there is no propagation in the y direction. The coordinate system is such that $\mathbf{r} = 0$ at some arbitrary point on the interface (where $z = 0$). The wavevectors for the incident (subscript 'i'), transmitted (subscript 't'), and reflected (subscript 'r') fields are:

$$\begin{aligned}\mathbf{k}_i &= \frac{\omega n_3}{c}(\hat{\mathbf{x}}\sin\theta_3 + \hat{\mathbf{z}}\cos\theta_3)\\ \mathbf{k}_t &= \frac{\omega n_1}{c}(\hat{\mathbf{x}}\sin\theta_1 + \hat{\mathbf{z}}\cos\theta_1)\\ \mathbf{k}_r &= \frac{\omega n_3}{c}(\hat{\mathbf{x}}\sin\theta_3 - \hat{\mathbf{z}}\cos\theta_3).\end{aligned} \tag{2.4}$$

For $n_3 > n_1$ and $\theta_3 > \theta_c$ only, the z component of wavevector $\mathbf{k}_t$ is imaginary (because of $\cos\theta_1$) and leads to the exponential decay of evanescence in the z-direction, as discussed below.

The incident field is a plane wave with the following forms (the superscripts p and s denote p-pol and s-pol, respectively):

$$\begin{aligned}\mathbf{E}_i^p &= E_{i0}^p e^{i(\mathbf{k}_i \cdot \mathbf{r} - \omega t)}\,(-\,\hat{\mathbf{x}}\cos\theta_3 + \hat{\mathbf{z}}\sin\theta_3)\\ \mathbf{E}_i^s &= E_{i0}^s e^{i(\mathbf{k}_i \cdot \mathbf{r} - \omega t)}\hat{\mathbf{y}}.\end{aligned} \tag{2.5}$$

Fresnel coefficients are the (possibly complex) factors by which the incident electric field amplitude at $\mathbf{r} = 0$ and $t = 0$ is multiplied to obtain the transmitted (possibly evanescent) field in medium 1 and the reflected field in medium 3 as a function of the incidence angle θ_3.

$$\begin{aligned}
&\textbf{\textit{transmitted:}}\\
&\quad \mathbf{E}_t^p = t_{31}^p E_{i0}^p e^{i(\mathbf{k}_t\cdot\mathbf{r}-\omega t)}\,(-\,\hat{\mathbf{x}}\cos\theta_1 + \hat{\mathbf{z}}\sin\theta_1)\\
&\quad \mathbf{E}_t^s = t_{31}^s E_{i0}^s e^{i(\mathbf{k}_t\cdot\mathbf{r}-\omega t)}\hat{\mathbf{y}}\\
&\textbf{\textit{reflected:}}\\
&\quad \mathbf{E}_r^p = r_{31}^s E_{i0}^p e^{i(\mathbf{k}_r\cdot\mathbf{r}-\omega t)}(\hat{\mathbf{x}}\cos\theta_3 + \hat{\mathbf{z}}\sin\theta_3)\\
&\quad \mathbf{E}_r^s = r_{31}^s E_{i0}^s e^{i(\mathbf{k}_r\cdot\mathbf{r}-\omega t)}\hat{\mathbf{y}}
\end{aligned} \tag{2.6}$$

where the Fresnel coefficients for light heading toward the interface from (in general) medium i to medium j are:

$$t_{ij}^p = \frac{2n_i\cos\theta_i}{n_i\cos\theta_j + n_j\cos\theta_i} \tag{2.7}$$

$$t_{ij}^s = \frac{2n_i\cos\theta_i}{n_i\cos\theta_i + n_j\cos\theta_j} \tag{2.8}$$

$$r_{ij}^p = \frac{n_j\cos\theta_i - n_i\cos\theta_j}{n_i\cos\theta_j + n_j\cos\theta_i} \tag{2.9}$$

$$r_{ij}^s = \frac{n_i\cos\theta_i - n_j\cos\theta_j}{n_i\cos\theta_i + n_j\cos\theta_j}. \tag{2.10}$$

The exponential decay in z for $\theta_3 > \theta_c$ comes from the complex exponential factor $e^{i\mathbf{k}_t\cdot\mathbf{r}}$ in $\mathbf{E}_t^{p,s}$ of equation (2.6), in which $\cos\theta_1$ becomes purely imaginary:

$$\begin{aligned}
e^{i\mathbf{k}_t\cdot\mathbf{r}} &= e^{\frac{i\omega n_1}{c}(\hat{\mathbf{x}}\sin\theta_1 + \hat{\mathbf{z}}\cos\theta_1)\cdot(x\hat{\mathbf{x}}+z\hat{\mathbf{z}})}\\
&= e^{\frac{i\omega n_3\sin\theta_3}{c}x}e^{\frac{-\omega n_1}{c}z\sqrt{(n_3/n_1)^2\sin^2\theta_3-1}}\\
&= e^{\frac{i\omega n_3\sin\theta_3}{c}x}e^{-\frac{z}{2d}},
\end{aligned} \tag{2.11}$$

where (written in several equivalent forms) the characteristic decay distance d is:

$$\begin{aligned}
d &= \frac{1}{\frac{2\omega n_1}{c}\sqrt{(n_3/n_1)^2\sin^2\theta_3 - 1}}\\
&= \frac{1}{2k_0 n_1\sqrt{(n_3/n_1)^2\sin^2\theta_3 - 1}}\\
&= \frac{\lambda_0}{4\pi n_1\sqrt{(n_3/n_1)^2\sin^2\theta_3 - 1}}\\
&= \frac{\lambda_0}{4\pi n_3\sqrt{\sin^2\theta_3 - \left(\frac{n_1}{n_3}\right)^2}}\\
&= \frac{\lambda_0}{4\pi n_3\sqrt{\sin^2\theta_3 - \sin^2\theta_c}}.
\end{aligned} \tag{2.12}$$

Note that d is the same for both polarizations. Except for supercritical $\theta_3 \to \theta_c$ (where $d \to \infty$), d is generally on the order of λ_0 or smaller.

Purely s-pol incident light (i.e. polarized in the y direction) produces a purely y-polarized evanescent field. However, p-pol incident light produces an evanescent field with a mix of z and x polarizations (and with an interesting phase relationship discussed below). This polarization mix is a direct consequence of the first of Maxwell's equations, which requires that the divergence of **E** must be zero everywhere, including in the evanescent field zone:

$$\nabla \cdot \mathbf{E} = \frac{\partial E_x}{\partial x} + \frac{\partial E_y}{\partial y} + \frac{\partial E_z}{\partial z} = 0. \tag{2.13}$$

For p-pol in our geometry, $Ey = 0$. If a p-pol evanescent field consisted purely of an exponentially decaying E_z in the z-direction with no E_x component. Since this would violate the requirement that the sum must be zero, then the last term in equation (2.13) would cycle between positive and negative (depending on the phase). But for the sum in equation (2.13) to be zero, the first term must exactly cancel the third term, proving that a nonzero E_x must be present in p-pol. However, if the evanescent field has infinite depth, as it does exactly at the critical angle, then the last term is zero, and there is no need for a non zero E_x.

How much of the intensity resides in the x component of p-pol evanescent light? The intensity of light is the square of the amplitude of the electric field. The ratio of the intensity of the x component vs the z component (easily derivable from equation (2.6)) for p-pol TIR is a function of the incidence angle. The ratio is $|-\cos\theta_1/\sin\theta_1|^2$. Recalling that $\cos\theta_1$ is purely imaginary for supercritical angles, the ratio can be rewritten as $1 - \sin^2\theta_c/\sin^2\theta_3$. Clearly, right at the critical angle $\theta_3 = \theta_c$, none of the intensity is polarized in the x direction. The ratio monotonically increases with θ_3 until it reaches $1 - \sin^2\theta_c = 1 - (n_1/n_3)^2$ at skimming (90°) incidence. For typical parameters in biochemistry and cell biology ($n_1 = 1.33$, $n_3 = 1.52$), this maximum ratio is 0.23. For the incidence angles that are more typically experimentally accessed using a TIR microscope setup, the ratio is ~0.04.

At subcritical (non-TIR) θ_3 incidence angles, both $\cos\theta_1$ and $\cos\theta_3$ are real, and consequently so are all the Fresnel coefficients. In the particular subcritical case of $n_3 > n_1$, the Fresnel coefficients are not only real but also all positive. This means that no new phase factors are introduced by the Fresnel coefficients, and the incident, transmitted, and reflected fields all have the same phase: i.e. the wavefront crests all match up with each other at the interface boundary, as shown for the incident and transmitted fields in figure 1.5(a). However, if $n_3 < n_1$ (a case of no particular relevance to TIR), the coefficients for reflection (the r ones) are negative, meaning that there is a 180° phase shift between incidence and reflection (crests matching up with troughs), as expected when light impinges upon an interface toward a higher refractive index.

On the other hand, for supercritical θ_3 angles (for which TIR occurs), all of the Fresnel coefficients are complex because of the purely imaginary nature of $\cos\theta_1$, and this introduces phase shifts such that the wavefront crests of incident, transmitted, and reflected fields do not match up at the interface. Figure 1.5(c) shows this mismatch. After some algebra, the Fresnel-coefficient-induced phase shifts of the evanescent (transmitted) field at the $z = 0$ interface, relative to the incident light, can be expressed as:

$$\delta_{pz}^{\text{evan}} = \tan^{-1}\left[-\frac{(\sin^2\theta_3 - \sin^2\theta_c)^{1/2}}{\sin^2\theta_c \cos\theta_3}\right] \tag{2.14}$$

$$\delta_{px}^{\text{evan}} = -\frac{\pi}{2} + \tan^{-1}\left[-\frac{(\sin^2\theta_3 - \sin^2\theta_c)^{1/2}}{\sin^2\theta_c \cos\theta_3}\right] \tag{2.15}$$

$$\delta_{sy}^{\text{evan}} = \tan^{-1}\left[-\frac{(\sin^2\theta_3 - \sin^2\theta_c)^{1/2}}{\cos\theta_3}\right]. \tag{2.16}$$

The phase shifts for p-pol (z component only) $\delta_{pz}^{\text{evan}}$ and for s-pol $\delta_{sy}^{\text{evan}}$ are such that an evanescent wavefront is shifted to the right (in the $+x$ direction) relative to where an incident wavefront impinges upon the surface. The phase shift varies smoothly from 0 radians at the critical angle to $-\pi/2$ radians at $\theta_3 = \pi/2$, and the most rapid rate of increase occurs in the few degrees above the critical angle.

An additional shift of $-\pi/2$ (moving wavefronts toward higher x) occurs in just the x-polarized component of p-pol $\delta_{px}^{\text{evan}}$, because of the $-\mathbf{x}\cos\theta_1$ that appears in equation (2.6a). This results in a unique feature of evanescent light: the z-polarized and phase-lagged x-polarized components add vectorially, to produce elliptical polarization in the evanescent field in the x–y travel plane of the wavefronts, in a kind of 'cartwheel.' By contrast, elliptical polarization in propagating light is always in a plane transverse to the propagation direction. The phase shifts (and elliptical polarization for p-pol) are schematically shown in figurc 2.1

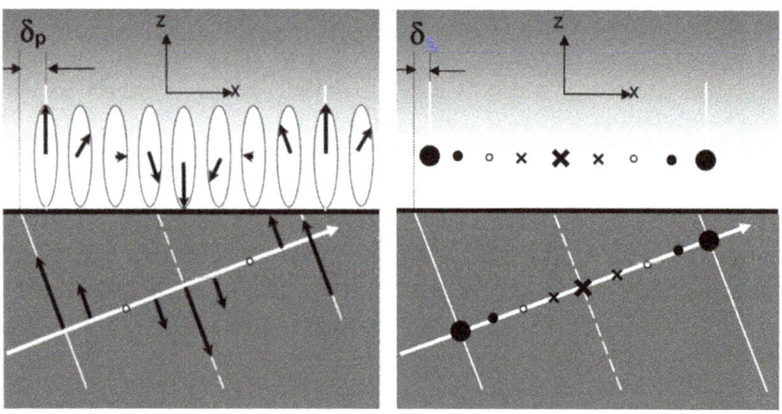

Figure 2.1. Polarization directions for p-pol (left) and s-pol (right) TIR. Reprinted from Axelrod (2008), copyright (2008), with permission from Elsevier.

In general, the phase shift for p-pol light is larger than that for s-pol light. Both approach zero as θ_3 approaches θ_c. The incidence-angle dependence of these phase shifts accounts for the 'Goos–Hänchen' effect (discussed in section 2.4), whereby a finite beam undergoing TIR 'appears' to propagate along the surface before reflecting back into medium 3.

Regardless of polarization, the complex exponential factor involving x in equation (2.11) represents the periodicity of wavefronts in the x direction in medium 1 along the surface, with a spacing λ_x:

$$\lambda_x = \frac{\lambda_0}{n_3 \sin \theta_3}. \tag{2.17}$$

where λ_0 is the light wavelength in vacuum ($n = 1$). Unlike the case of freely propagating light, the evanescent spatial period is not at all affected by medium 1 in which it resides. It is determined only by the spacing of the incident light wavefronts in medium 3 as they intersect the interface. This spacing can be experimentally important, because it determines the spacing of interference fringes produced when two coherent TIR beams illuminate the same region of the surface from different directions (as used in structured TIR illumination, to be discussed in chapter 5). This spacing can be made quite small by using a material 3 with a high refractive index.

Full expressions for the evanescent fields can be rewritten from equation (2.6) as follows:

$$\mathbf{E}_t^p(z) = 2E_{i0}^p \cos\theta_3(\sin^4\theta_c \cos^2\theta_3 + \sin^2\theta_3 - \sin^2\theta_c)^{-1/2}$$
$$e^{i\delta_p}[-i(\sin^2\theta_3 - \sin^2\theta_c)^{1/2}\,\hat{\mathbf{x}} + \sin\theta_3\hat{\mathbf{z}}]e^{i(k_x x-\omega t)}e^{-z/2d}$$

$$\mathbf{E}_t^s(z) = 2E_{i0}^s\frac{\cos\theta_3}{\cos\theta_c}e^{i\delta_s}\hat{\mathbf{y}}e^{i(k_x x-\omega t)}e^{-z/2d}. \tag{2.18}$$

2.3 Evanescent intensity

The transmitted intensities $I_1^{p,s}$ are proportional to the scalar product $\mathbf{E}_t^* \cdot \mathbf{E}_t$. From equation (2.18), we get:

$$I_1^{p,s}(z, \theta_3) = I_1^{p,s}(0, \theta_3)e^{-z/d}. \tag{2.19}$$

Since the vector factors in equation (2.6) are all of unit amplitude, the intensities at $z = 0$ (normalized to incident intensity) are simply the amplitude-squared Fresnel coefficients from equations (2.7) and (2.8):

$$I_1^{p,s}(0) = |t_{31}^{p,s}|^2 = (t_{31}^{p,s})(t_{31}^{p,s})^*, \tag{2.20}$$

where the asterisk indicates a complex conjugate. $I^{p,s}$ is proportional to the probability of excitation of a fluorophore residing at any point in the evanescent field space. (This statement, however, needs some qualification. It is true: (a) assuming the applicability of first-order time-dependent perturbation theory, in which the fluorophore is treated as a quantum system but the light is treated classically; (b) in the limit at which the light is dim enough that multiphoton effects are not observed; and (c) if the fluorophore is initially in the ground state.)

$I^{p,s}$ can be divided into two regimes, one for subcritical angles $\theta_3 < \theta_c$ and one for supercritical angles (evanescent, $\theta_3 > \theta_c$), After some straightforward algebra, we obtain

subcritical

$$I_1^p(0) = \frac{(2\cos\theta_3)^2 \sin^2\theta_c}{(\sin^2\theta_c - \sin^2\theta_3) + (\sin^4\theta_c)(\cos\theta_3)^2 + 2\sin^2\theta_c\sqrt{\sin^2\theta_c - \sin^2\theta_3}}$$
$$I_1^s(0) = \frac{(2\cos\theta_3)^2}{\left(\cos\theta_3 + \sqrt{\sin^2\theta_c - \sin^2\theta_3}\right)^2} \tag{2.21}$$

supercritical

$$I_1^p(0) = \frac{(2\cos\theta_3)^2 \sin^2\theta_c}{(\sin^2\theta_3 - \sin^2\theta_c) + \sin^4\theta_c \cos^2\theta_3}$$
$$I_1^s(0) = \frac{(2\cos\theta_3)^2}{\cos^2\theta_c}. \tag{2.22}$$

Figure 2.2 plots $I_1^{p,s}(0)$ as a function of θ_3 for both their sub- and supercritical angular ranges. The evanescent intensity $I_1^p(0)$ is slightly greater than $I_1^s(0)$ at all supercritical angles. Both reach a maximum at precisely the critical angle, and both decrease monotonically to zero at the extreme incidence angle of $\theta_3 = \pi/2$. Note that

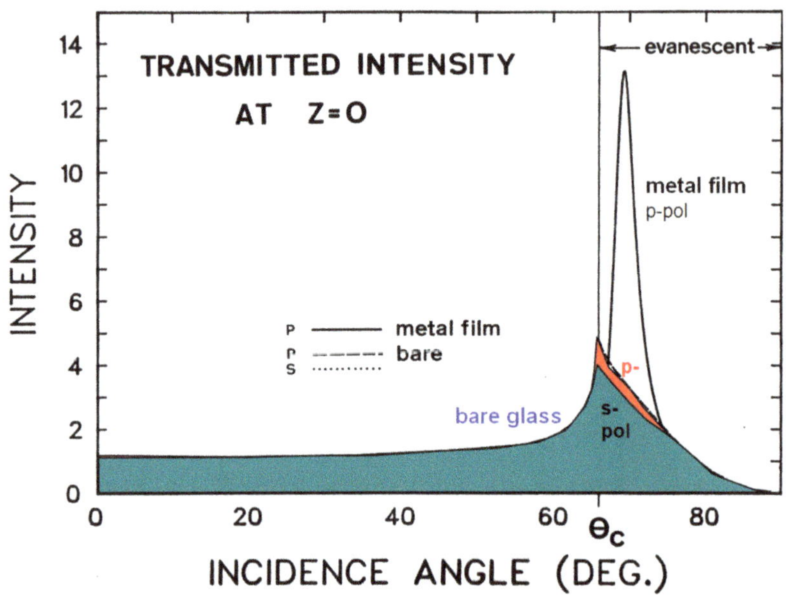

Figure 2.2. Transmitted intensity vs incidence angle θ_3. The prominent narrow peak that occurs in the supercritical range only occurs for glass coated with a metal film. This is a 'surface plasmon' effect, discussed in section 3.3. Reprinted from Axelrod (2008), copyright (2008), with permission from Elsevier.

the p-pol intensity $I_1^p(0)$ includes the contributions of both its x- and z-polarized components.

This graph depicts a couple of surprising features. The subcritical angle intensity of the transmitted wave is slightly *greater* than the incident intensity, even at normal incidence ($\theta_3 = 0$); one might expect slightly *less* intensity because of reflection at the surface. The discrepancy at $\theta_3 = 0$ arises because the intensity in figure 2.3 refers to the squared amplitude of the electric field (EE^*) alone, rather than to the actual energy flux of the light (the Poynting vector), which involves the product of EE^* and the refractive index of the medium in which the light propagates.

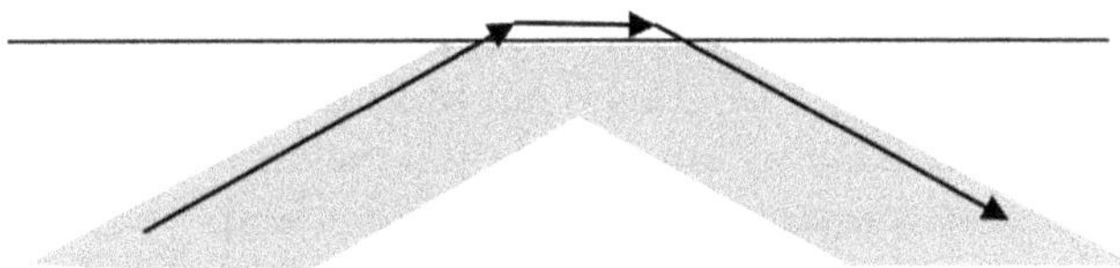

Figure 2.3. Schematic view of energy flow across the TIR interface for a finite-width beam.

Another surprising features is that the supercritical angle evanescent intensity maximum is exactly four times the incident intensity at the critical angle for s-pol, and even slightly higher for p-pol. Without any absorption in the system, all of the energy simply returns back into medium 3 in the reflected beam. However, if the energy of this enhanced intensity could somehow be withdrawn from the system, it might seem to violate conservation of energy. In most circumstances in fluorescence microscopy, this seeming contradiction is not a pressing matter. If a fluorophore is used as a 'probe' of intensity, we assume that the fraction of available energy absorbed and re-emitted by a fluorophore used to probe this intensity is very small. Under this assumption, this intensity enhancement does not violate conservation of energy because the fluorophore is assumed not to drain the evanescent field energy by itself.

Nonetheless, the probability of absorption by each fluorophore probe is still up to four times as likely in an evanescent field than in the incident field that generated it (for s-pol, and slightly more for p pol). Given enough fluorophore, the evanescent field intensity thereby obtained would be decremented by fewer fluorophore probes than would be the case in the incident field. The fluorophore probe can be viewed in the same way as a 'load' in a simple voltage-source electrical circuit, where a small load (high resistance) does not affect the voltage but a large load (low resistance) does. For example, consider a voltage source in the form of a step-up AC electrical transformer, where the output (secondary) voltage may be, say, four times higher than the input (primary) voltage. The output power can be no greater than the input power; the output voltage is merely less robust (more easily decremented) by a current 'load.'

These topics are related to energy flow in an evanescent field. One can show that there is no net flow of energy in the z-direction from medium 3 to medium 1 once the evanescent field is established (Born and Wolf (1999), section 1.5.3). However, the evanescent field does contain electromagnetic energy; how did it get there?

This energy migrated across the interface in the first few femtoseconds when the incident light was first turned on. Similarly, if the incident field had a finite width (say, a laser beam), Poynting vector calculations would show that energy does indeed flow across the boundary near one edge of the finite beam, then travels along the interface in medium 1, then returns across the interface back into medium 3 at the other edge of the finite beam (see figure 2.3).

Calculations of energy flow across the boundary must take into account not only electric field amplitudes, but also the Poynting vector's ability to cross a unit area of surface, which depends on the obliquity of its angle with respect to the surface.

Although the emission from a fluorophore excited by an evanescent field as viewed by a microscopic objective might be expected to follow an exponential decay with z according to equation (2.18), this is not precisely true. Fluorescence emission near a dielectric interface is rather anisotropic and the degree of anisotropy is itself z dependent. One cause of this anisotropy of emission is simply partial reflection from the interface and consequent optical interference between the direct and reflected emission beams. However, another more subtle cause is the interaction of the fluorophore's 'near field' with the interface and its consequent conversion into light propagating at high angles into the higher-index n_3 material (the solid substrate). This subject is treated in detail in chapter 4. In general, the closer a fluorophore is to the surface, the larger the proportion of its emitted light that enters the substrate. If a sufficiently high-aperture objective is used to gather the near-field emission through the substrate, then the effective thickness of the surface detection zone is reduced beyond the surface selectivity generated by the evanescent field excitation of TIR alone. Low-aperture objectives are not as sensitive to this enhanced effect, because they fail to capture the near-field light. On its own, the near-field emission capture effect can be the basis of surface-selective fluorescence detection and imaging.

2.4 Finite-width incident beams: the Goos–Hänchen shift

For a TIR Gaussian laser beam focused with a typically narrow angle of convergence (producing a finite-width, nearly collimated beam), the experimentally observed evanescent illumination is approximately an elliptical Gaussian profile, and the polarization and penetration depth are approximately equal to those of a single infinite plane wave. But there is a complication. The evanescent wave can be pictured as the beam's partial emergence from the solid into the liquid, its journey along the surface for some finite distance, and then re-entrance into the solid. (A cute anthropomorphic description was presented as 'The photon emerges into the low-density medium, discovers it does not want to be there, and then dives back.') The distance of travel along the surface is measurable for a finite-width beam and is called the Goos–Hänchen shift. The Goos–Hänchen shift ranges monotonically from infinite at $\theta = \theta_c$ (which of course corresponds to the refracted beam skimming along the interface) to a fraction of a wavelength at $\theta_3 = \pi/2$. In this way, TIR is not a sudden discontinuous effect, but a smooth transition from an infinite to a finite skimming distance as the supercritical incidence angle increases.

Any finite-width propagating incident beam can be expressed mathematically as an integral of infinite plane waves approaching at a range of incidence angles, each with an appropriate phase. Referred to as the 'angular spectrum of plane waves,' it is a direct consequence of the Fourier transform of the finite beam's field. The physical reason that the integral over infinite plane waves appears to have a finite width is that the phases of the plane waves 'match up' near the center of the finite beam, leading to constructive interference. Conversely, (because of the range of propagation angles), the plane waves are out-of-step at the periphery of the finite beam and beyond, leading to destructive interference. (see figure 2.4)

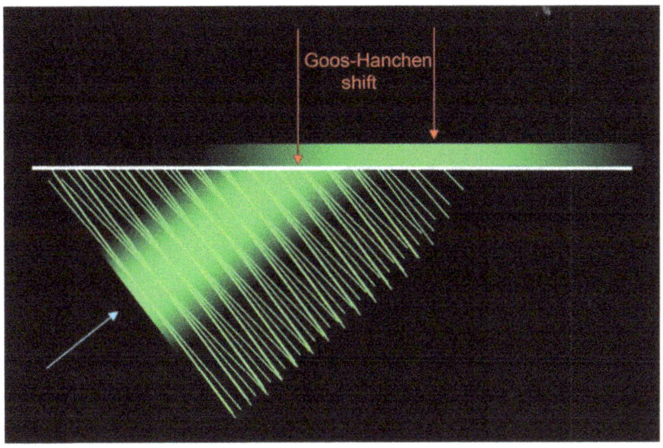

Figure 2.4. Angular spectrum of plane waves creates a finite-width incident beam. The intensities are depicted as lighter green; the evanescent intensity is for $z = 0$ only.

The Goos–Hänchen effect is a shift (to the right in figure 2.4, exaggerated for pictorial clarity there) of the brightest region of the evanescent field. It is a direct consequence of the angular spectrum of the incident beam: specifically, the increasing evanescent phase shift (equations (2.14) and (2.15)); and the shortening of the wavefront spacing in the x direction ($\lambda_x = 2\pi/k_x$) as the incidence angle of each component plane wave increases.

Figure 2.5 shows this effect qualitatively. Consider just two component plane waves of the incident beam, one at the critical angle (shown in blue) and the other at a slightly higher angle (shown in red). The location of the brightest part of the incident beam (where the blue and red wavefronts constructively interfere) is shown as a black arrow. Wherever a blue wavefront intersects the surface, it gives rise to a blue evanescent wavefront with no phase shift (because the phase shift is zero for critical angle incidence). However, the red evanescent wavefronts are shifted to the right (to more positive x) as predicted for supercritical light by equations (2.14) and (2.15). This places the red wavefronts to the right of the blue wavefronts, at least those near the central position of the incoming beam. However, the spacing of the red wavefronts is slightly smaller than that of the blue ones because of the oblique geometry, so the red wavefronts progressively 'lose ground' to the blue as x increases, until they match. At this spot (indicated by the dashed black arrow),

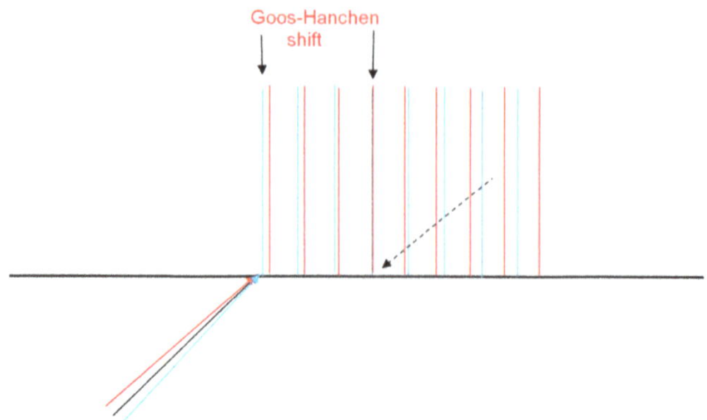

Figure 2.5. Schematic view of the Goos–Hänchen shift arising from the incidence-angle dependence of the evanescent field phase shift. An actual incident beam consists of a smooth spectrum of plane-wave angles, all of which only have the same phase at the beam's central position. Here, only two plane waves in that spectrum are shown.

the evanescent field is the brightest. Farther toward the $(+x)$ right, the red wavefronts continue to fall behind the blue ones and the evanescent field becomes dimmer. The result is that the brightest part of the evanescent field is shifted to the right of the brightest part of the incident beam. This is the Goos–Hänchen shift. The Goos–Hänchen shift is slightly different in amplitude for the s polarization than for the p-polarization because their phase shifts vs incidence angle (equations (2.14) and (2.15)) are slightly different.

2.5 Reflected intensities

Up to this point, we have concentrated on evanescent intensities, as found in medium 1. In general, the reflected intensities $I_3^{p,s}$ are proportional to the scalar product $\mathbf{E}_r^* \cdot \mathbf{E}_r$. In the simple case of a single interface (no intermediate film), equation (2.6) shows that the reflected intensity (normalized to the incident intensity) is

$$I_3^{p,s}(0) = |r_{31}^{p,s}|^2 = (r_{31}^{p,s})(r_{31}^{p,s})^*, \tag{2.23}$$

with the reflection Fresnel coefficients given by equations (2.9) and (2.10). When the incidence angle passes from subcritical to supercritical, $\cos\theta_1$ flips from purely real to purely imaginary. In the subcritical zone, $|r_{31}|^2 < 1$. In the supercritical zone, r_{31} has the form $(a - ib)/(a + ib)$, so $|r_{31}|^2$ equals exactly unity, regardless of polarization (where the dummy variables a and b are both real). This, of course, is where the 'total' in the term 'total internal reflection' originates. In the presence of an intermediate film (medium 2), the reflected intensities, as found in medium 3, can exhibit nontrivial behavior. This subject will be covered in the next chapter.

In the expression for reflected fields (equation (2.6)) rather than intensities, the possibly complex Fresnel reflection coefficient r_{31} appears to the first power. For subcritical incidence angles (and assuming $n_3 > n_1$), r_{31} is entirely real and so the reflected electric field has zero phase shift relative to the incident light. This means that the incident beam and the reflected beam wavefronts 'match up' where they touch the interface. For supercritical angles, however, as $\cos\theta_1$ becomes imaginary, r_{31} becomes complex, and so the reflected field has a nonzero phase shift relative to the incident light. This means that the incident beam and the reflected beam wavefronts do not match up where they touch the interface, and, in fact, the reflected wavefronts are shifted in the $+x$ direction. After applying some algebra to r_{31}, the reflected light phase shifts can be shown to be:

$$\delta_p^{\text{refl}} = \tan^{-1}\frac{-2n_1^2n_3^2\cos\theta_3(\sin^2\theta_3 - \sin^2\theta_c)^{1/2}}{n_1^4\cos^2\theta_3 - n_3^4(\sin^2\theta_3 - \sin^2\theta_c)} \tag{2.24}$$

$$\delta_s^{\text{refl}} = \tan^{-1}\frac{-2\cos\theta_3(\sin^2\theta_3 - \sin^2\theta_c)^{1/2}}{\cos^2\theta_3 - (\sin^2\theta_3 - \sin^2\theta_c)}. \tag{2.25}$$

Recall that for the evanescent field, there are two different phase lags for the p-pol, one for x and one for z polarization, thereby leading to elliptical polarization. However, for reflection, there is a single phase lag for both the x and z components upon reflection. This means that linear p-pol incident light remains linearly polarized in p-pol upon total reflection.

Figure 2.6 shows the phase lags for both polarizations, for both reflection and evanescent fields.

In addition, the difference in phase lag between s-pol and p-pol reflected light means that incident light, linearly polarized at 45° to the plane of incidence (so that

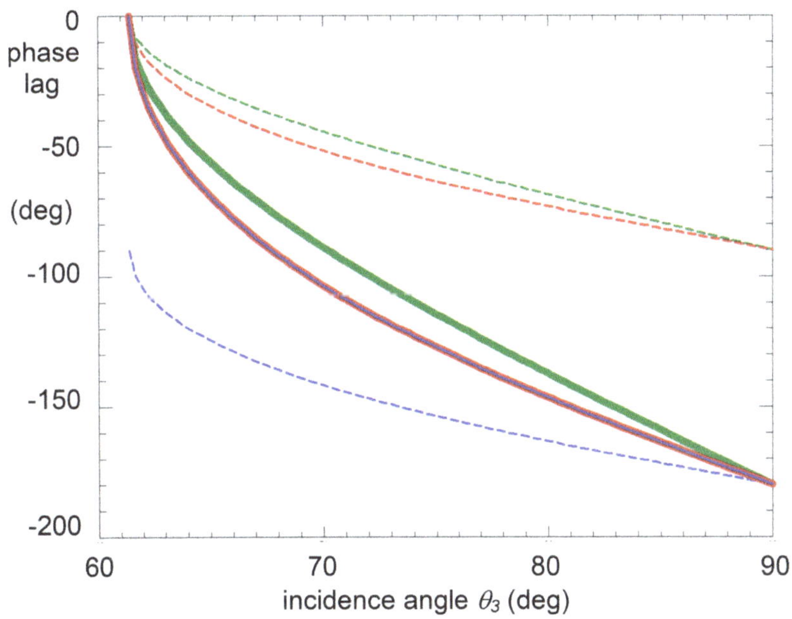

Figure 2.6. Phase lags (in degrees) vs supercritical incidence angle for evanescent (dashed) and reflected (solid) fields, and for the polarizations s-pol (green), z polarization (red), and x polarization (blue), for refractive indices $n_1 = 1.33$ and $n_3 = 1.515$.

its p- and s-pol components are in phase and have equal amplitude) is totally reflected while becoming elliptically polarized. This ellipticity is small, but it is a sensitive function of the refractive index on the low-density side of the interface and should be detectable as a measure of this refractive index (see figure 2.7).

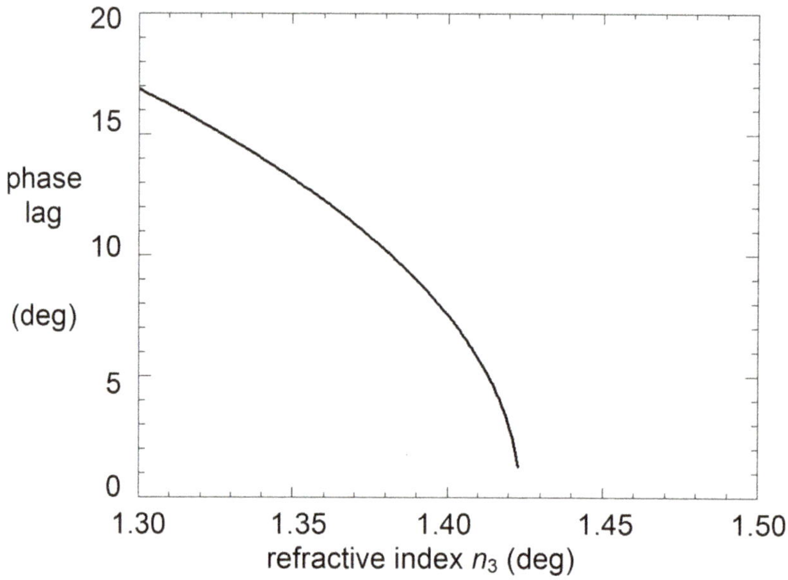

Figure 2.7. Phase difference between s-pol and p-pol (s minus p) reflected beams vs refractive index n_1, for the single supercritical angle $\theta_3 = 70°$ and $n_3 = 1.515$.

Note that the phase delay for the reflected field is exactly double that for the evanescent field. In other words, the evanescent wavefronts intersect the interface exactly halfway between the incident and reflected wavefronts, as shown in figure 2.8.

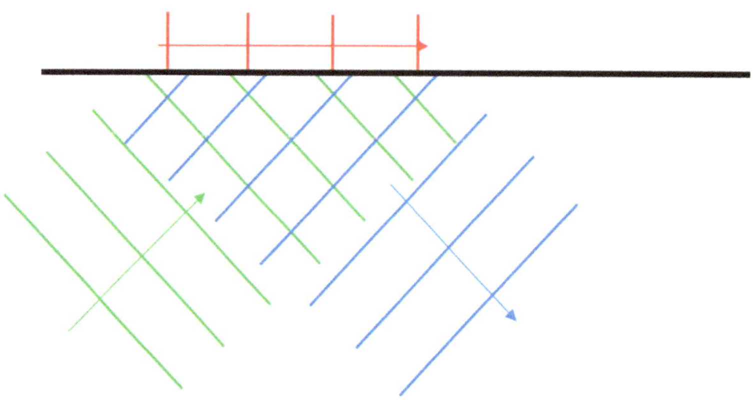

Figure 2.8. Relative positions of wavefronts in TIR. Note that the evanescent wavefronts (shown in red) are halfway between the incident (green) and reflected (blue) wavefronts.

Further reading

Born E and Wolf E 1999 *Principles of Optics: Electromagnetic Theory of Propagation, Interference and Diffraction of Light* 7th edn (Cambridge: Cambridge University Press) ch 1

Axelrod D 2008 Total internal reflection fluorescence microscopy *Methods in Cell Biology: Biophysical Tools for Biologists 2: Methods In Vivo* ed J J Correia and H W Detrich (New York: Academic Press) pp 169–221

Chapter 3

Structure in the lower-index material

Section 3.5 of this chapter is coauthored with Jeremy J Axelrod, Department of Physics, University of California, Berkeley.

Chapter 2 presented the fundamental principles of total internal reflection (TIR), but it was limited to a single-interface case in which both media were entirely uniform and non-absorptive. More common circumstances, however, involve an n_1 which contains an imaginary part (representing light absorption by a chromophore or fluorophore), varies in the z direction (including the case in which an intermediate layer, perhaps metal, is interposed between medium 1 and medium 3, leading to surface plasmon effects), or varies in the x–y direction (which leads to scattering). This chapter covers those situations.

3.1 Light absorption in medium 1

If medium 1 is capable of absorbing light (say, by containing some solubilized chromophore), then n_1 contains an additional unitless imaginary term κ:

$$\hat{n}_1 = n_1 + i\kappa. \tag{3.1}$$

The parameter κ is proportional to the concentration of the chromophore and to its intrinsic ability to absorb photons. By inserting the complex form (equation 3.1) wherever the purely real n_1 appears in the basic theory for fields (i.e. in equations 2.3, 2.4, and 2.6), the reflected and transmitted light intensities and the z-profile at the interface above the critical TIR critical angle can be tracked as a function of κ.

But what range of κ is relevant? To draw a connection between κ and the more familiar parameters for light absorption used in physical chemistry, consider the case of normal incidence ($\theta_3 = \theta_1 = 0$). The z-dependence of the transmitted field then becomes

$$e^{i\mathbf{k}_t \cdot \mathbf{r}} = e^{\frac{i\omega\hat{n}_1 \cos\theta_1}{c}z} = e^{i\frac{2\pi}{\lambda}\hat{n}_1 z} = e^{i\frac{2\pi}{\lambda}n_1 z}e^{-\frac{2\pi}{\lambda}\kappa z}. \tag{3.2}$$

The corresponding intensity then has a real exponential decay in z, which is usually written in the base-ten form involving 'optical density' A:

$$e^{-\frac{4\pi}{\lambda}\kappa z} = 10^{-A}, \tag{3.3}$$

where $A = \varepsilon Cl$, ε is the chromophore's intrinsic 'extinction coefficient' in units of mol · L^{-1} · cm^{-1}); C is the chromophore's concentration in mol · L^{-1}; and l is the depth of solution traversed by the light in units of cm. Therefore,

$$\frac{4\pi}{\lambda \ln 10}\kappa z = \varepsilon Cl. \tag{3.4}$$

Taking care to express the vacuum wavelength λ in cm so that $l = z$, we get

$$\kappa = \frac{\varepsilon C\lambda \ln 10}{4\pi}. \tag{3.5}$$

The practical range of κ in TIRF microscopy experiments can now be roughly determined. In TIRF microscopy, it is possible to detect fluorophore numbers in the range of 1 to 1,000,000 within the evanescent field depth of a lateral resolution area (a volume of roughly 100 nm × 200 nm × 200 nm). Converting this into the equivalent moles/liter for C, plugging that result into equation (3.5), and approximating $\varepsilon = 100{,}000$ liters/(mole-cm) for typical fluorophores, κ can be calculated to vary over the range of about 10^{-9} to 10^{-3} for typical TIRF applications.

Figure 3.1(a) shows the reflected intensity vs incidence angle. The black curve (zero-absorption case) rapidly reaches 1.0 at the TIR critical angle at a sharp corner and remains pinned there for supercritical angles, as expected for TIR. However, further increasing the κ values smooths the curve and essentially abolishes the 'totality' of TIR at supercritical angles. (The highest two κ values are considerably beyond the range of values likely to be encountered with standard molecular fluorophore labels in TIRF microscopy.)

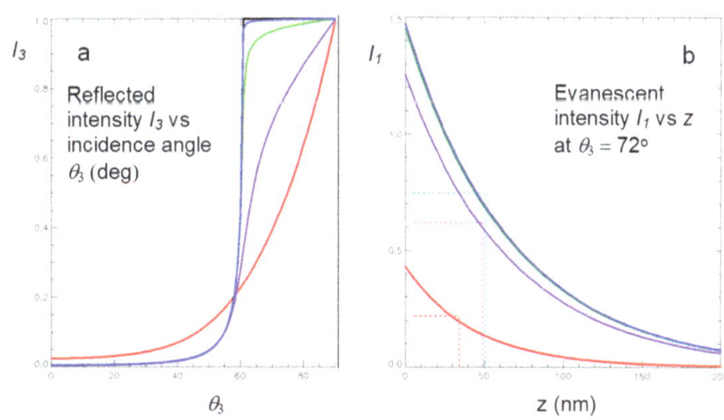

Figure 3.1. (a) The θ_3 dependence of the reflected intensity (in medium 3) for a range of κ values = 0.0 (black), 0.0004 (blue), 0.004 (green), 0.04 (purple), and 0.4 (red). The incident intensity is set to unity. (b) The evanescent intensity (in medium 1) vs z measured at an incidence angle of $\theta_3 = 72°$, for the same range of κ values.

Figure 3.1(b) shows the z-dependence of the evanescent intensity for the same range of κ values. The three lowest values (which cover the entire range of practical TIRF microscopy experiments) are virtually superimposed. In such experiments, absorption barely affects the evanescent decay rate at all. The intensity of the evanescent field is also not significantly affected. Only at the highest κ values is the profile of the evanescent field affected, and the main effect there is on the intensity. Higher κ values do decrease the decay distance slightly (as indicated by the dashed lines at half-decay), but it is a fairly small effect.

The most important practical conclusion is that both the evanescent intensity and depth remain unaltered from the zero-absorption case for all fluorophore concentrations likely to be encountered in TIRF microscopy experiments.

3.2 Intermediate layers

In actual experiments in biophysics, chemistry, or cell biology, the interface may not be a simple interface between two media, but rather a stratified multilayer system. Examples include: a collagen coating, a biological membrane, or a lipid bilayer interposed between glass and an aqueous medium. Another example is a thin metal-film coating, which can be used to quench fluorescence within the first ~10 nm of the surface (see section 6.3). Also for a metal film, at a certain sharply defined angle of incidence θ_p ('the surface plasmon angle'), the p-polarized evanescent intensity becomes an order of magnitude brighter than the incident light at the peak (as seen in figure 2.2). The topic of surface plasmons will be discussed in section 3.3.

We discuss here the TIR evanescent wave in a three-layer system (figure 3.2) in which incident light travels from medium 3 (refractive index n_3) through an intermediate layer (n_2) of thickness h toward medium 1 (n_1). Snell's law is still applicable here:

$$n_1 \sin \theta_1 = n_2 \sin \theta_2 = n_3 \sin \theta_3 \quad (3.6)$$

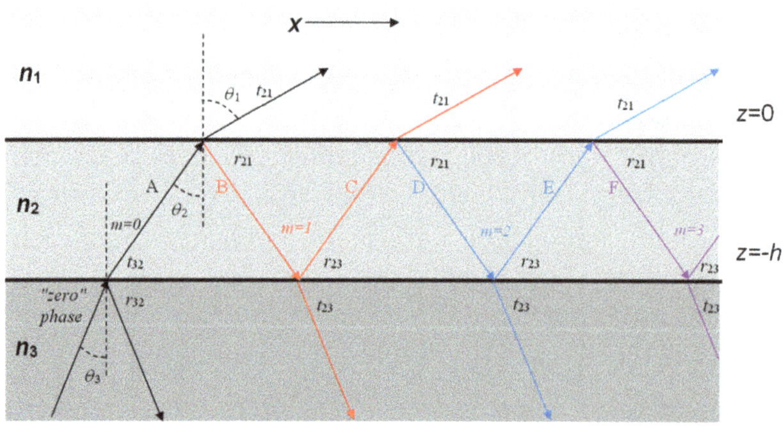

Figure 3.2. Three-layer system shown in ray-optics form, with each pair of ray reflections shown in a different color.

as is the trigonometric identity in any medium $i = 1, 2$, or 3:

$$\cos\theta_i = \sqrt{1 - \sin^2\theta_i}. \tag{3.7}$$

Evanescence can still occur within medium $i = 2$ or 3 (or both), provided that Snell's law makes $\sin\theta_i > 1$ and the corresponding $\cos\theta_i$ purely imaginary.

Some important qualitative features can be noted first, followed by the quantitative math.

- The insertion of an intermediate layer never thwarts TIR if it would occur between n_3 and n_1 at some supercritical angle θ_3, regardless of the intermediate layer's refractive index n_2 or thickness. The only question is whether TIR takes place at the n_3:n_2 interface or the n_2:n_1 interface. Since the intermediate layer is likely to be very thin (no deeper than several tens of nanometers) in many applications, precisely which interface supports TIR is not important for qualitative studies.
- Regardless of n_2 and the thickness of the intermediate layer, the evanescent wave's profile in medium 1 decays exponentially with a characteristic decay distance given by equation (2.12). However, the intermediate layer affects its intensity at the interface with medium 1 and the overall distance of penetration of the field as measured from the surface of medium 3.

In the three-medium/two-interface case considered here (see figure 3.2), the incident plane wave refracts from medium 3 into medium 2 and then necessarily experiences multiple reflections in medium 2. These are infinite plane waves, but the direction of propagation of each is shown as a solid line (a 'beam'), starting from some arbitrary x position where the phase is defined at be 'zero.' For each 'bounce' at an interface with medium 1 or medium 3, a Fresnel coefficient describes the amplitude and phase shift for the reflection (r) or transmission (t) that occurs there. The light also accumulates phase shifts simply by traversing some distance in each medium. The phase shift (in radians per distance of travel in the z-direction component) is denoted by β_i, where

$$\beta_i = (2\pi n_i \cos\theta_i)/\lambda_0 \tag{3.8}$$

and λ_0 is the light wavelength in vacuum. If either $\cos\theta_1$ or $\cos\theta_2$ is imaginary (corresponding to evanescent rather than propagating light in the corresponding medium), then so too is the corresponding β_1 or β_2. In the case of evanescence, the geometrical picture of propagation does not apply literally, but the mathematical formalism still does.

The sum of all these electric field contributions at any point is a convergent geometric series, in which each term in medium 2 represents one leg of the zig-zag path with its associated amplitude loss and accumulating phase shift, or each beam in medium 1 or medium 3 that is transmitted away. In general, a geometric series can be written in closed form as

$$\sum_{m=0}^{\infty} \alpha^m = \frac{1}{1 - \alpha}, \tag{3.9}$$

where α is a (possibly complex) parameter with $|\alpha| < 1$.

Once the electric field vectors are calculated, the intensity in the p-pol and s-pol polarizations can be calculated from the x, y, and z components of the fields:

$$\begin{aligned} I_i^p(z, \theta_3) &= |E_{ix}|^2 + |E_{iz}|^2 \\ I_i^s(z, \theta_3) &= |E_{iy}|^2. \end{aligned} \tag{3.10}$$

3.2.1 Field and intensity in medium 1 ($z \geq 0$)

To set up the transmitted field in medium 1, the light must first be transmitted through the 3:2 interface. Then it hits the 2:1 interface, partially transmits there, and then partially reflects repeatedly in a zig-zag pattern, first 'bouncing' off the 2:1 interface, then the 3:2 interface. With each reflection at 2:1, it partially transmits again into medium 1. With each pair of reflections, the light also traverses the thickness h of the film twice (at an oblique angle θ_2), thereby experiencing a phase shift by a factor of $\exp(2i\beta_2 h)$.

Once the light beam escapes into medium 1, it experiences an additional phase shift factor of $\exp(i\beta_1 z)$ as it proceeds along $z > 0$. Consequently, counting the reflection pairs by integer $m = 1, 2, \ldots$, the field amplitude in medium 1 at $z = 0$ can be written in general as:

$$E_t(z, \theta_3) = t_{32} e^{i\beta_2 h} t_{21} e^{i\beta_1 z} \sum_{m=0}^{\infty} (r_{21} r_{23} e^{2i\beta_2 h})^m = t_{32} t_{21} e^{i\beta_2 h} e^{i\beta_1 z} \left(\frac{1}{1 - r_{21} r_{23} e^{2i\beta_2 h}} \right), \tag{3.11}$$

where t_{ij} and r_{ij} are the single-interface transmission and reflection Fresnel coefficients shown in equations (2.7–2.10). Concerning the order of the subscripts, the light beam is assumed to be heading *from* medium i (= 1, 2, or 3) *toward* medium j.

The fields in medium 1 resulting from each leg of the 'bounce' all have the same direction and polarization, therefore, including the polarization is straightforward:

$$\mathbf{E}_t^p(z) = \frac{t_{32}^p t_{21}^p (e^{i\beta_2 h} e^{i\beta_1 z})}{1 - r_{21}^p r_{23}^p e^{2i\beta_2 h}} (-\hat{\mathbf{x}} \cos\theta_1 + \hat{\mathbf{z}} \sin\theta_1) \tag{3.12}$$

$$\mathbf{E}_t^s(z) = \frac{t_{32}^s t_{21}^s e^{i\beta_2 h} e^{i\beta_1 z}}{1 - r_{21}^s r_{23}^s e^{2i\beta_2 h}} \hat{\mathbf{y}}. \tag{3.13}$$

Based on these expressions, the intensity in medium 1 can be written as:

$$I_t^{p,s}(z, \theta_3) = \left| \frac{t_{32}^{p,s} t_{21}^{p,s} (e^{i\beta_2 h} e^{i\beta_1 z})}{1 - r_{21}^{p,s} r_{23}^{p,s} e^{2i\beta_2 h}} \right|^2 e^{-z/d}. \tag{3.14}$$

Despite the complications related to the angles, the film's refractive index, and its thickness implied by the field equations (2.26–27), the z-dependence of the evanescent intensity in medium 1 exponentially decays with exactly the same characteristic distance d as it would for a single interface (with the same n_1 and n_3); see equation (2.14). This characteristic distance does not depend at all upon the film thickness h or the refractive index n_2.

The intensity $I_1^{p,s}(0, \theta_3)$ at the immediate surface of the film in medium 1 is particularly interesting in the intermediate zone between two critical angles, $\theta_{31,c} < \theta_3 < \theta_{32,c}$, in which the incident light propagates within the film and is evanescent in medium 1. In this notation, $\theta_{ij,c}$ is the critical angle for light at an interface between medium j and medium i, i.e.

$$\theta_{ij,c} = \sin^{-1}\frac{n_j}{n_1}. \tag{3.15}$$

Figure 3.3 shows $I_1^s(0, \theta_3)$ (normalized to the incident intensity) vs θ_3 for various h, for a typical set of refractive indices that might be encountered in biochemical or cell biological samples. The p-pol $I_1^p(0, \theta_3)$ is not shown but is qualitatively similar. The lower graph is just a continuation of the upper graph for a higher range of h.

For the case of no film ($h = 0$), the evanescent intensity slowly decreases monotonically with increasing θ_3, reaching exactly zero at $\theta_3° = 90°$. However, in the presence of a film ($h > 0$), the evanescent intensity $I_1^s(0, \theta_3)$ shows a strong dependence on h, reaching maxima at particular θ_3 values in the intermediate zone that can exceed the evanescent intensity in the $h = 0$ (no-film) case by a factor of more than two. This resonance-like behavior results from multiple reflections in the film. It is analogous to the classic effects of an oil film on water or an antireflection coating on glass, except that here we are looking at the amplitude of an evanescent field rather than a reflection of propagating light. The exact conditions for constructive or destructive interference are complicated, because the phase shifts at the medium 2:medium 1 TIR interface are neither exactly zero nor exactly 180° but are instead intermediate, depending upon θ_2, n_1, and the polarization.

Several practical consequences arise from the effects depicted in figure 3.3:

- The existence of a film (which otherwise might not have been apparent) should be obvious from the shape of $I_1^s(0, \theta_3)$ vs θ_3 within the $\theta_{31,c} < \theta_3 < \theta_{32,c}$ range. This intensity can be reported by emission from a fluorophore in medium 1 adsorbed to medium 2's surface. For thin films (e.g. with $h = 0.1$), $I_1^s(0, \theta_3)$ monotonically decreases, but its decrease is significantly more rapid than in the no-film case. For thick films ($h > 0.6\lambda$), the $I_1^s(0, \theta_3)$ curves are no longer monotonic decays and show at least one clear maximum in the $\theta_3 > \theta_{31,c}$ range, which is qualitatively different from the no-film case.
- For thicker films ($h > 0.6\lambda$), the thickness range should be readable from the number of maxima, and (if the film's refractive index is known) the exact thickness can be deduced from the location of the maxima in $I_1^s(0, \theta_3)$.
- A film can boost the evanescent intensity in medium 1 by at least a factor of two over its no-film value at the same incidence angle for a judicious choice of θ_3. This evanescent enhancement may be useful where laser power is marginal or where more rapid fluorophore photobleaching is desired.

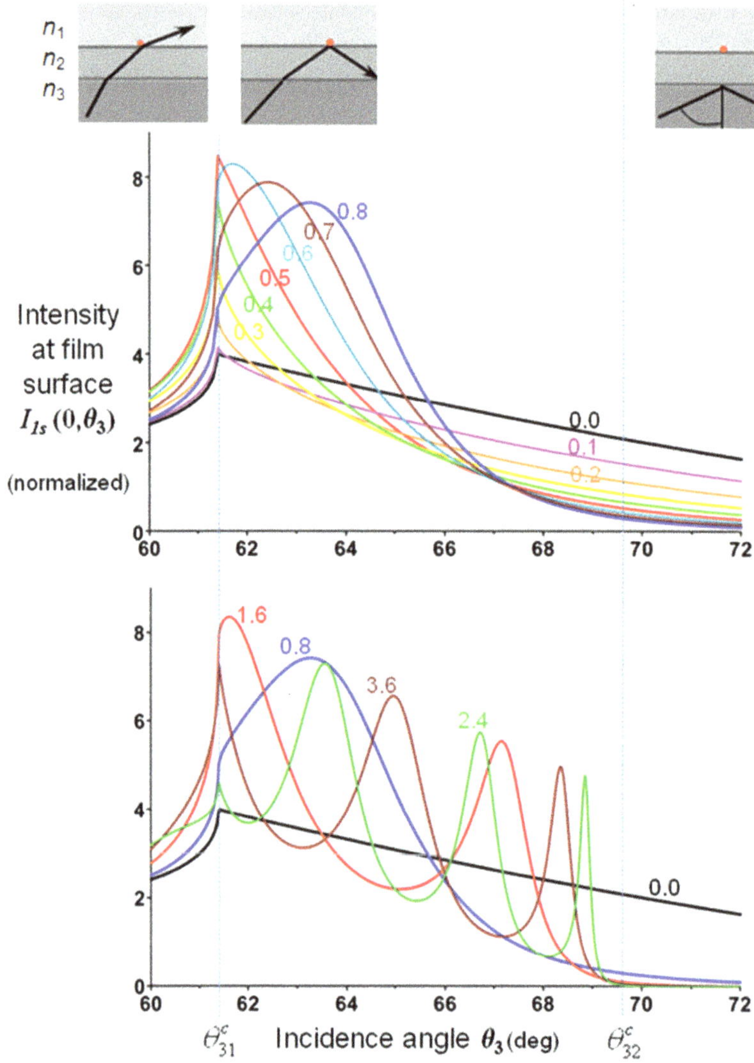

Figure 3.3. Intensity of light at the n_1:n_2 (z = 0) interface as a function of incidence angle θ_3 for various color-coded film thicknesses h. The intensities are normalized to the incident intensity; the thicknesses h are normalized to the vacuum wavelength λ_o. Three distinct regimes are shown: no TIR (left); TIR at n_1:n_2 (middle); and TIR at n_2:n_3, as shown in the drawings at the top. Only s-pol is shown. The bottom panel is the same as the upper panel, except that it displays a range of larger h values. The refractive indices assumed are: n_1 = 1.33; n_2 = 1.42; and n_3 = 1.515. Reprinted from Axelrod (2012), copyright (2012) John Wiley & Sons.

- The effects are qualitatively similar for both p-pol and s-pol incident light (only s-pol is shown in figure 3.3). The only difference is a very slightly lower contrast between maxima and minima and a very slight shift (~0.1°) of the maxima to smaller angles in the p-pol case.
- If an experimental measurement of $I_1^s(0, \theta_3)$ does not match any curves typified by those in figure 3.3 (including the h = 0 case), one can strongly

conclude that sample surface does indeed support a film and that it must be heterogeneous in thickness and/or refractive index.

- The $I_1^{p,s}(0, \theta_3)$ curves are very sensitive to variations in n_1 for higher film thicknesses. This sensitivity may not be interesting in most cases because n_1 is typically known and is also homogeneous. However, if the 2:1 interface is overlaid with a heterogeneous structure (such as adsorbed cells), then different lateral locations in the sample exhibit locally different $I_1^{p,s}(0, \theta_3)$ curves. Scanning θ_3 while recording wide-field images should reveal a new contrast mechanism based on the absolute value of n_1 rather than gradients in n_1.
- The $I_1^{p,s}(0, \theta_3)$ curves are also very sensitive to the film's refractive index n_2, especially for higher film thicknesses, for which these curves are non-monotonic. For example, decreasing n_2 by only 0.01 from 1.42 to 1.41 (at a constant $h = 2\lambda_0$) shifts both prominent maxima of $I_1^s(0, \theta_3)$ to lower θ_3 values by about 1°. Figure 3.4 shows this shift in detail. This effect is not due to a simple change in the optical path length of the film, which is proportional to the product $n_2 h$. A decrease in n_2 can be distinguished from a similar-factored increase in h; these two parameters affect $I_1^s(0, \theta_3)$ differently. Therefore, it should be possible to curve-fit a theoretical $I_1^s(0, \theta_3)$ to experimental data to unambiguously determine both n_2 and h simultaneously.

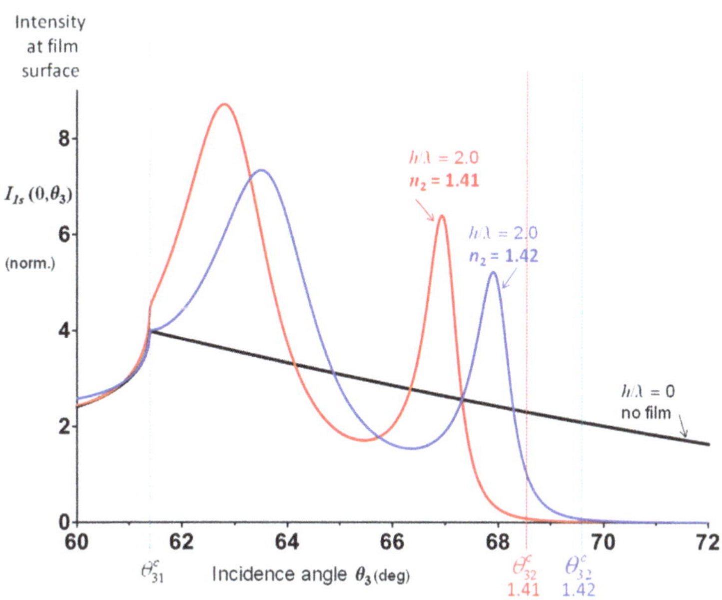

Figure 3.4. Intensity I_1 at $z = 0$ in medium 1 vs incidence angle θ_3 for two (slightly) different film refractive indices $n_2 = 1.41$ (red) and $n_2 = 1.42$ (blue). In both cases, $h = 2.0$ (normalized to the vacuum wavelength), $n_1 = 1.33$, and $n_3 = 1.515$. The no-film case ($h = 0$, black) is also shown. Reprinted from Axelrod (2012), copyright (2012) John Wiley & Sons.

3.2.2 Field and intensity in medium 2 ($-h < z < 0$)

The field within the intermediate layer is an infinite geometric sum of 'upward' oriented and 'downward' oriented legs in its zig-zag path as depicted 'on the page' in figure 3.2. For s-pol, the fields in the 'up' and 'down' legs all have the same vectorial direction (**y**), i.e. perpendicularly 'into the page' in figure 3.2. But for p-pol, the fields in the 'up' legs have a different direction than the fields in the 'down' legs. The results are:

$$\begin{aligned}\mathbf{E}^p_{\text{film}} &= t^p_{32}e^{i\beta_2 h}[e^{i\beta_2 z}(-\hat{\mathbf{x}}\cos\theta_2 + \hat{\mathbf{z}}\sin\theta_2) \\ &\quad + r^p_{21}e^{-i\beta_2 z}(\hat{\mathbf{x}}\cos\theta_2 + \hat{\mathbf{z}}\sin\theta_2)]\left[\frac{1}{1 - r^p_{21}r^p_{23}e^{2i\beta_2 h}}\right] \\ \mathbf{E}^s_{\text{film}} &= t^s_{32}e^{i\beta_2 h}(e^{i\beta_2 z} + r^s_{21}e^{-i\beta_2 z})\left[\frac{1}{1 - r^s_{21}r^s_{23}e^{2i\beta_2 h}}\right]\hat{\mathbf{y}}.\end{aligned} \tag{3.16}$$

To show the intensity inside the film, figure 3.5 displays intensity information as grey-scale brightness on a 2D (θ_3,z) plane. Boundary conditions imposed by Maxwell's equations lead to continuity (regardless of the refractive-index ratio) of E_x and E_y across the 2:1 ($z = 0$) interface, and a ratio of $(n_2/n_1)^2$ in E_z in traversing from medium 2 to medium 1. For particular θ_3, z, and h/λ_0 combinations, the brightest intensities are found deeper within the film. If h/λ_0 and n_2 are known, then the shape of $I_1^{p,s}(z < 0, \theta_3)$ can indicate the z-position of a fluorophore embedded in medium 2.

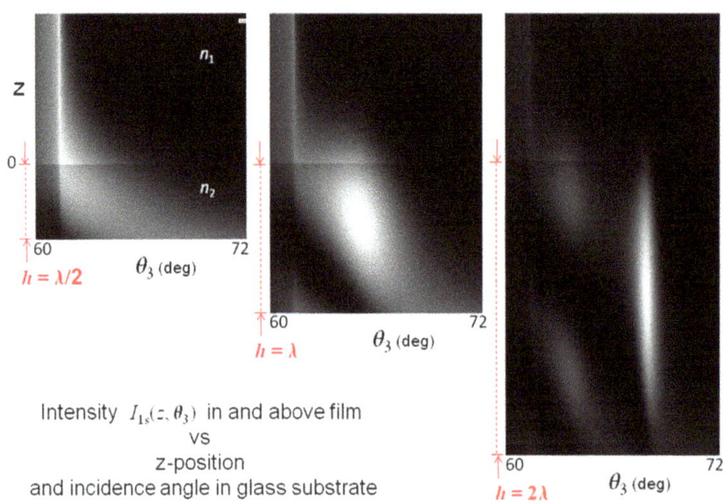

Figure 3.5. Intensities in medium 1 ($h > 0$) and medium 2 ($h < 0$) vs z as the incidence angle θ_3 is increased from sub- to supercritical for three different film thicknesses. Reprinted from Axelrod (2012), copyright (2012) John Wiley & Sons.

3.2.3 Field and intensity in medium 3 ($z < -h$)

The reflected light behavior as a function of incidence angle θ_3 is of experimental interest as a means to study the refractive indices in medium 3 and in the film (and the film thickness). The electric fields in medium 3 are:

$$\mathbf{E}_r^p = \left[r_{32}^s + t_{32}^s t_{23}^s r_{21}^s e^{2i\beta_2 h}\left(\frac{1}{1 - r_{23}^s r_{21}^s e^{2i\beta_2 h}}\right)\right](\hat{\mathbf{x}}\cos\theta_1 + \hat{\mathbf{z}}\sin\theta_1)$$
$$\mathbf{E}_r^s = \left[r_{32}^s + t_{32}^s t_{23}^s r_{21}^s e^{2i\beta_2 h}\left(\frac{1}{1 - r_{23}^s r_{21}^s e^{2i\beta_2 h}}\right)\right]\hat{\mathbf{y}}. \tag{3.17}$$

The above rather complicated-looking term in the square brackets can be rewritten using the following identities that are derived directly from the expressions for the Fresnel coefficients equations (2.7–2.10):

$$r_{23} = -r_{32}$$
and
$$t_{32}t_{23} = 1 - r_{32}^2,$$

so that

$$\mathbf{E}_r^p = \left(\frac{r_{32}^s + r_{21}^s e^{2i\beta_2 h}}{1 + r_{32}^s r_{21}^s e^{2i\beta_2 h}}\right)(\hat{\mathbf{x}}\cos\theta_1 + \hat{\mathbf{z}}\sin\theta_1)$$
$$\mathbf{E}_r^s = \left(\frac{r_{32}^s + r_{21}^s e^{2i\beta_2 h}}{1 + r_{32}^s r_{21}^s e^{2i\beta_2 h}}\right)\hat{\mathbf{y}}. \tag{3.18}$$

Consequently, the intensity of the reflected light at a three-layer surface is

$$I_3^{p,s}(\theta_3) = \left|\frac{r_{32}^{p,s} + r_{21}^{p,s} e^{2i\beta_2 h}}{1 + r_{32}^{p,s} r_{21}^{p,s} e^{2i\beta_2 h}}\right|^2. \tag{3.19}$$

This reflection expression is the 'partner' to the evanescent intensity equation (3.14) in medium 1, taking into account the multiple reflections in medium 2, and notably does not contain a z-dependence.

For unpolarized or circularly polarized incident light, the intensity is just the average of the p-pol and s-pol I values as calculated above.

Figure 3.6 shows the theoretical I_3 vs θ_3 for a range of n_1 values, for $h = 0$ only, valid for either polarization. Each curve in this set of curves shows a steep rise up to the TIR critical angle at which I_3 achieves a value of one; the effect of increasing n_1 is mainly a shift toward higher θ_3 while maintaining the steep shape. Of course, with $h = 0$, it follows that $\beta_2 = 0$, and I_3 vs θ_3 is independent of the value of n_2.

Figure 3.7 shows I vs θ_3 for a particular single n_1 but with a range of intermediate layer h/λ values. The effect of $h > 0$ is to broaden I vs θ_3 so that it starts rising at smaller θ_3 while still achieving TIR at the same critical angle. The clear distinction in I_3 between the different effects of n_1 and h upon their variation vs θ_3 can be the basis for distinguishing n_1 and h. Of course, if $n_2 \sim n_1$, then the term 'intermediate layer' has little meaning, and varying h/λ_0 has little effect.

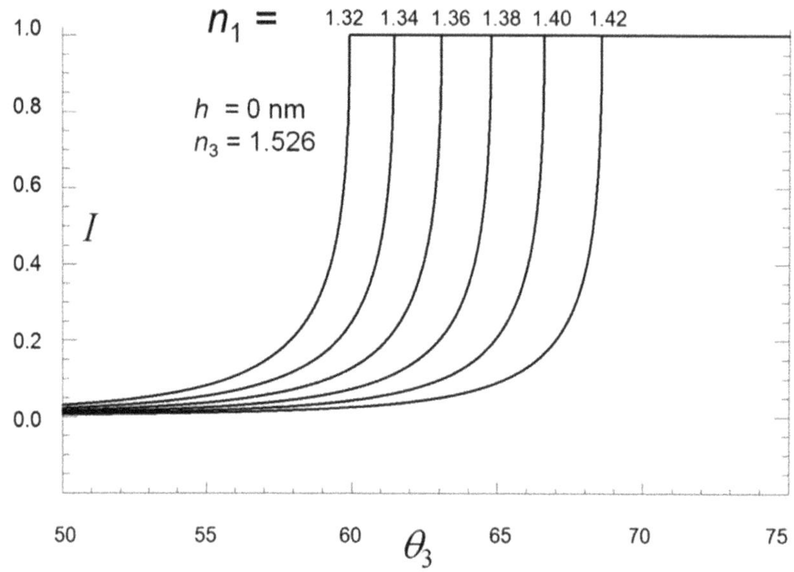

Figure 3.6. Intensity of the reflected beam in medium 3 vs incidence angle for several different n_1 values. Reprinted from Bohannon *et al* (2017), copyright Microscopy Society of America 2017.

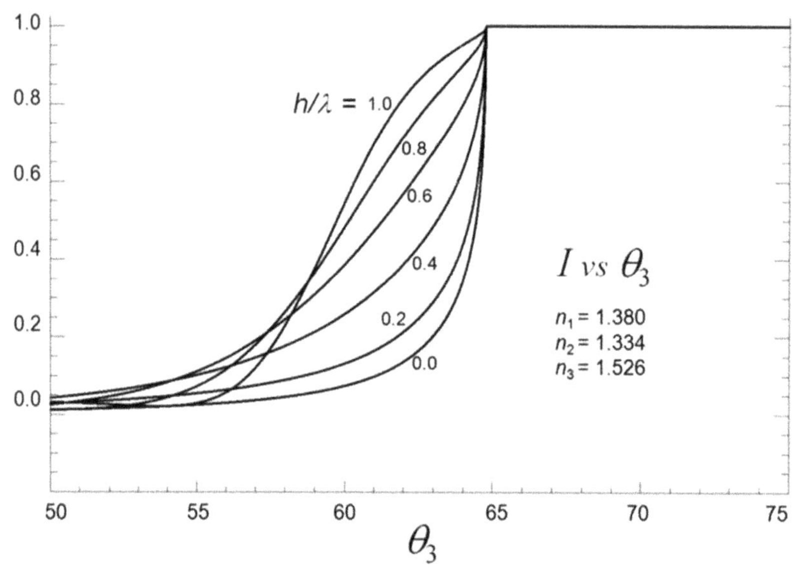

Figure 3.7. Intensity of reflected beam in medium 3 vs incidence angle for several different film thicknesses (h values). Reprinted from Bohannon *et al* (2017), copyright Microscopy Society of America 2017.

3.3 Metal films and surface plasmons

A striking behavior appears if the film is a metal: the evanescent wave intensity in medium 1 can reach a sharp peak at some incidence angle greater than the critical angle, called a 'surface plasmon resonance' peak. This is depicted in figure 2.2.

For nonconductive transparent materials, the dielectric constant ε (which is the square of the refractive index) is largely real and positive. However, for metals at visible optical frequencies, ε has a large negative real part (plus a smaller imaginary part corresponding to energy loss due to a nonzero electrical resistance). For a metal, therefore, the refractive index n_2 (the square root of the dielectric constant) is largely imaginary. The exponential decay rate in medium 1 is unaffected by whether medium 2 is a metal or a nonconductor because Snell's law (equation (2.20)) that relates θ_3 to θ_1 remains valid. The main effect is on the amplitude of the evanescent field.

Mathematically, the factor t_{21} in the numerator of I_t (equation (2.28)) is responsible for the sharp evanescent intensity peak because of a divide-by-zero singularity. For p-pol, the t_{21} denominator is

$$\begin{aligned} & n_2 \cos\theta_1 + n_1 \cos\theta_2 \\ &= n_2 \cos\theta_1 + n_1\sqrt{1 - \sin^2\theta_2} \\ &= n_2 \cos\theta_1 + n_1\sqrt{1 - \left(\frac{n_3}{n_2}\right)^2 \sin^2\theta_3}\,. \end{aligned} \tag{3.20}$$

Since n_2 for a metal is largely imaginary (and ignoring for now its relatively small real part), the square-root term is real and positive. Thus this whole denominator expression can be zero only if the first term $n_2 \cos\theta_1$ is real and negative. Since $\cos\theta_1$ (for evanescence) is imaginary, the term $n_2 \cos\theta_1$ is indeed real and negative, and the whole denominator can be zero if:

$$n_2 \cos\theta_1 + n_1\sqrt{1 - \left(\frac{n_3}{n_2}\right)^2 \sin^2\theta_3} = 0. \tag{3.21}$$

This can be solved for the incidence angle $\theta_3^{\max}$ at which I_t reaches a maximum. Using relative dielectric constants $\varepsilon_i = n_i^2$, straightforward algebra yields:

$$\theta_3^{\max} = \sin^{-1}\left[\left(\frac{\varepsilon_3}{\varepsilon_1} + \frac{\varepsilon_3}{\varepsilon_2}\right)^{-1/2}\right]. \tag{3.22}$$

Some parameters likely to be encountered with common materials and $\lambda_0 - 520$ nm are: $\varepsilon_1 = 1.77$ (water), $\varepsilon_2 = -32.5 + 8.4i$ (aluminum), and $\varepsilon_3 = 2.31$ (glass). Then $\theta_3^{\max} = 65°$. This is slightly larger than the critical angle of 61° at which the evanescent field would be the brightest were there no film at all. For an aluminum, gold, or silver film at a glass–water interface, $\theta_3^{\max}$ is greater than the critical angle θ_c for TIR. The intensity enhancement is rather remarkable, since a 20 nm-thick (or more) metal film is almost opaque to the eye but still supports an evanescent field much stronger than would exist in the absence of a film.

The discussion above applies to p-pol. For s-pol, the subscripts are reshuffled, so the t_{21} denominator is:

$$n_2 \cos\theta_2 + n_1 \cos\theta_1 = n_2\sqrt{1 - \left(\frac{n_3}{n_2}\right)^2 \sin^2\theta_3} + n_1 \cos\theta_1. \tag{3.23}$$

In this s-pol case, the square-root term is again real and positive. However, here, it multiplies a largely imaginary (and large-amplitude) n_2, so the product cannot possibly cancel the $n_1 \cos \theta_1$ term, which is real and positive. In fact, the denominator's magnitude is quite large at all incidence angles, meaning that the metal film reduces the s-polarized evanescent intensity to nearly zero at all incidence angles. Therefore, the evanescent field just outside a metal film is almost completely p-polarized.

Physically, the strongly peaked evanescent intensity effect is due to a resonant excitation of electron oscillations at the metal–water interface. This resonance can be understood qualitatively. The metal is a uniform sea of free electrons. If a group of them are displaced, leaving a nonuniform distribution, an electronic 'restoring force,' based on their charge, will try to restore their equilibrium distribution in the sea. But the electrons do have a mass, so in returning to their equilibrium distribution, they experience an inertial 'overshoot.' This behavior then implies a 'natural frequency' of oscillation in the metal. If an oscillating driving force, in the form of an external optical EM field, matches this natural frequency, a resonance occurs, allowing the electron oscillations to become large. This condition is only optimally met if the driving field has a properly tuned wavelength spacing and frequency.

There are some potentially useful experimental consequences of TIR excitation through a thin metal film coated on glass:

- The metal film almost totally quenches fluorescence within the first 10 nm of the surface, and the quenching effect is virtually absent at a distance of 100 nm (a matter to be discussed in chapter 4). Therefore, TIR with a metal-film coated glass can be used to selectively excite fluorophores in the 10–200 nm distance range.
- A light beam incident upon a 20 nm aluminum film from the glass side at a glass–aluminum film–water interface evidently does not have to be collimated to produce TIR. Those rays that are incident at the surface plasmon angle create a strong evanescent wave; those rays that have an incidence angle that is too low or high create a negligible field in the water. This phenomenon may ease the practical requirement for a collimated incident beam in TIR and make it easier to set up TIR with a conventional arc light source.
- The surface plasmon angle is strongly sensitive to the index of refraction on the low-density side of the interface. In a sample that has a spatially varying refractive index (such as an adhered cell), some areas support the surface plasmon resonance while neighboring areas do not. The locations supporting resonance show much stronger light scattering. This effect can be used as the basis of 'surface plasmon microscopy.'
- The metal film leads to a highly polarized evanescent wave, regardless of the purity of the incident polarization.

3.4 Slab waveguides

Optical fibers and many chemical sensors depend upon a propagating light wave being 'trapped' in high-index material, small in one or two dimensions, which is

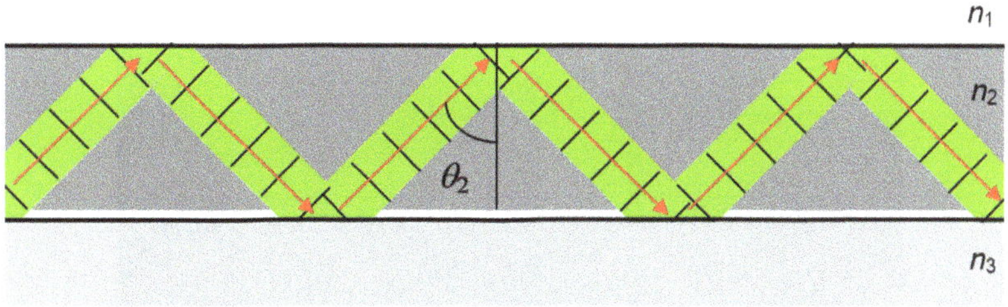

Figure 3.8. Multiple total internal reflections in a slab where the beam width is less than the slab thickness.

unable to refract out because of TIR at its interfaces, where it contacts lower-refractive-index materials. This subject deserves (and has received) whole books on its own. However, some of its fundamental principles can be seen even in the simplified case of a 'slab waveguide,' which is just a special case of the planar intermediate layer theory discussed above, in particular, the case of $n_1 = n_3 < n_2$

Figure 3.8 shows a qualitative view of a 1D slab waveguide for the case in which the beam width is less than the separation between total reflections.

In this case, the beam does not overlap or interfere with itself, and apart from the requirement that TIR occurs, there is no restriction on the angle θ_2, or the thickness h that can support longitudinal propagation (in the $+x$ direction, i.e. to the right) in the slab.

The opposite extreme, in which the beam width is 'infinite,' each reflection interferes with previous reflections. This the case for which equation (3.16) applies directly. In general, the wavefronts from the many reflections are out of phase and add up to zero intensity. However, for just the right combination of angle, film thickness, and phase delay at each reflection, the wavefronts are in phase and produce a strong intensity. $\mathbf{E}_{\text{film}}^{p,s}$ contains a factor of the form

$$\left[\frac{1}{1 - (r^{p,s})^2 e^{2i\beta_2 h}}\right], \tag{3.24}$$

which 'explodes' if the denominator is zero. This can occur only for very specific combinations of θ_2 and h. Figure 3.9 shows a semilog grey scale plot of the intensity of light trapped inside the film as averaged over the z-thickness of the film.

Several interesting features are depicted here.

- As a semilog grey scale, the figure compresses a huge range of intensities varying over ten orders of magnitude from the darkest to the brightest places. At very particular sharply defined 'resonant' pairs of (θ_2, h) where the figure image is the brightest, the film is essentially transparent in the x direction and light can appear at very distant $+x$ positions, whereas at any other pairs, it is virtually opaque.
- Only supercritical values of θ_2 are plotted here, because at subcritical angles, the energy is not trapped in the film. However, this raises the question: how can light be introduced into the film? Propagating light that simply refracts in

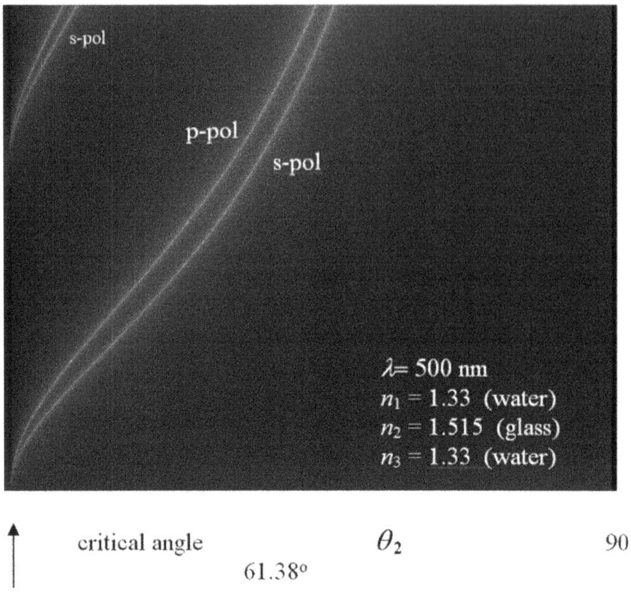

Figure 3.9. Intensity in the film $\left\langle \left| \mathbf{E}_{film}^{p,s} \right|^2 \right\rangle$ vs film thickness h and angle θ_2. Brackets (⟨⟩) denote an average over the z-depth of the film.

from the 1:2 or the 2:3 interface will not produce a supercritical θ_2. A specially designed prism or diffraction grating in contact with an interface, or a focusing system that introduces light in through the end, can be used.

- For very small h, no light can be propagated in the x direction in the film, regardless of the value of θ_2.
- The exact coordinates of a resonant (θ_2,h) pair depend not only upon geometry, but also upon the θ_2-dependent phase shift that occurs with each total reflection This is the same nontrivial phase shift that gives rise to the Goos–Hänchen shift discussed earlier. This phase shift is also dependent on the refractive indices inside and outside the film.
- This phase shift is also polarization dependent, which then leads to a very distinct set of resonant (θ_2,h) pairs for p-pol vs s-pol. This means that light that does propagate in the film toward distant $+x$ positions is highly polarized, even if the input light was unpolarized.
- The sharpness of the resonant peak at some h for any particular θ_2 should be exquisitely sensitive to slight changes in h or in the refractive indices inside or outside the film. Experimental determination of the location of the resonant (θ_2,h) pair should provide a direct measure of the thickness h of the film.
- The larger the film thickness, the larger the number of (still countable) angles that lead to resonance at that thickness. Note the new set of bright lines that appear at high h in the upper left of figure 3.9.

Of course, a 1D slab film, thin in z but infinite in x and y, is not a practical device. However, the physics leading to discrete resonance combinations is still relevant for devices that are thin in 2D, such as near-field scanning optical probes.

Optical waveguides are particularly useful as biosensors. The surface of a waveguide (say, a glass containing silicon dioxide) can be chemically functionalized with immobilized target molecules (by chemistry involving organosilanes) to bind specifically to certain constituents in an unknown fluid mix. This binding can provide an optical signal by at least three general methods: (a) *Surface proximity.* If the constituent is fluorescent, then it will 'light up' only when it is in the evanescent field at the waveguide surface. (b) *Conformation change.* A constituent binding may change the conformation of the immobilized target and allow it to convert from fluorescent to non-fluorescent (or vice-versa). (c) *Energy transfer.* The proximity of the constituent and the target may lead to energy transfer from one to the other and a consequent change in fluorescence, either brightening or quenching.

TIR slab waveguides can be very sensitive as biosensors for specific bindings that induce fluorescence. As will be discussed in the next chapter, the near-field evanescent emission of a fluorophore proximal to a surface is converted to propagating light in the slab at an angle that exceeds the critical angle. The fluorescence is thereby 'trapped' between the opposite walls of the slab and emerges only at the ends. The slab essentially harvests the fluorescence from a large surface area and concentrates it at its end, where a photodetector monitors it with a high degree of sensitivity.

An optical variation of a TIRF slab waveguide allows excitation light to be introduced and trapped in a slab without the need for an external prism. A glass (or plastic) surface can be etched or embossed with closely spaced grooves that behave like a diffraction grating, so that normally incident light is diffracted away from the normal at the surface. Coated upon the grooved glass is a thin intermediate film of a material (such as titanium dioxide) that has a refractive index higher than that of either the glass substrate or the water at its boundaries. The incident light diffracted into the intermediate film has a sufficient angle not only to totally reflect first at the thin film–aqueous interface but also to totally reflect at the subsequent thin film–glass interface, thereby trapping the light in the film. At a very particular angle (which depends on the wavelength and the film thickness), the film acts like an optical waveguide in which the reflections at the two surfaces create wavefronts that constructively add in phase with each other. The result is a very strong field in the high-index film and a corresponding very strong evanescent field in the aqueous medium (which surrounds the whole system) that can be used to excite fluorescence in biosensor devices. The thin-film waveguide surface preparation should also be applicable to the observation of living cells in a microscope.

Another application of a TIR slab waveguide has significant commercial and medical diagnosis possibilities for the simultaneous bioassay of many constituents. The surface of a substrate can be divided into a regular array of tiny, specially functionalized, chemically specific regions. The non-glass binding 'other end' of the organosilanes can be derivatized locally by a target molecule, a different one for each tiny region. The whole slab (while supporting TIR) would be immersed in a fluid containing an unknown mix of numerous interesting constituents. Specific binding in each region would then 'turn on' or 'turn off' the fluorescence in that region, depending on whether the constituent matched to the target were present in the unknown fluid mix. The TIRF patterns of possibly hundreds of regions could be monitored in one field by microscope imaging.

3.5 Total internal reflection scattering

So far, we have assumed that the TIR interface is smooth and laterally homogeneous. This produces an evanescent field that has the same z-dependent intensity everywhere in the x–y plane. Actual interfaces may contain refractive-index inhomogeneities $n_1(x,y,z)$, particularly if TIR is used to study living cells or liposomes adhered to the TIR surface. If these inhomogeneities were only gradually varying over large ($\gg\lambda$) distances in the x–y plane, then the evanescent field intensity and profile could still be calculated at each x–y location using the appropriate local value of n_1. However, in general, the inhomogeneities may vary sharply over much smaller distances, and then theory becomes more complicated. The inhomogeneities produce scattering of the evanescent field, some of which propagates away from the surface and thereby reduces the nominal surface selectivity of TIR. This section presents a theory for predicting the intensity and pattern of the altered field in medium 1. The result predicts a strong dependence on the wavelength, incidence angle θ_3, the spatial pattern of the refractive-index variations, as well as a (somewhat weaker) dependence on polarization. The goal here is to quantitatively predict how much surface selectivity is maintained, particularly for living cell cultures, given any specific input spatial pattern of refractive-index variations.

First, a note on notation. All of the theory in this section refers to electric fields in medium 1, so carrying the subscript '1' to distinguish it from media 2 and 3 is unnecessary. Instead, the subscript '1' applied to an electric field, in this section only, refers to scattered light. The subscript '0' applied to an electric field refers to unscattered incident evanescent light.

We will calculate the vector field of the scattered light $\mathbf{E}_1$ as generated by an evanescent incident field $\mathbf{E}_0$, but using a significant first-order approximation: that $|\mathbf{E}_1| << |\mathbf{E}_0|$. In this approximation, any scattered light that subsequently rescatters on refractive-index inhomogeneities is much less significant than the first scattering, i.e. we can ignore multiple scattering. On the other hand, if it were to turn out that $|\mathbf{E}_1| \sim |\mathbf{E}_0|$, then multiple scattering would be a distinct possibility, and the first-order approximation used here would be inadequate.

Consider a lower-index dielectric medium n_1 that contains small spatial variations in refractive index. These variations disrupt the evanescent field and produce propagating scattering. See figure 3.10 for a diagram of the geometry and the redefined variables.

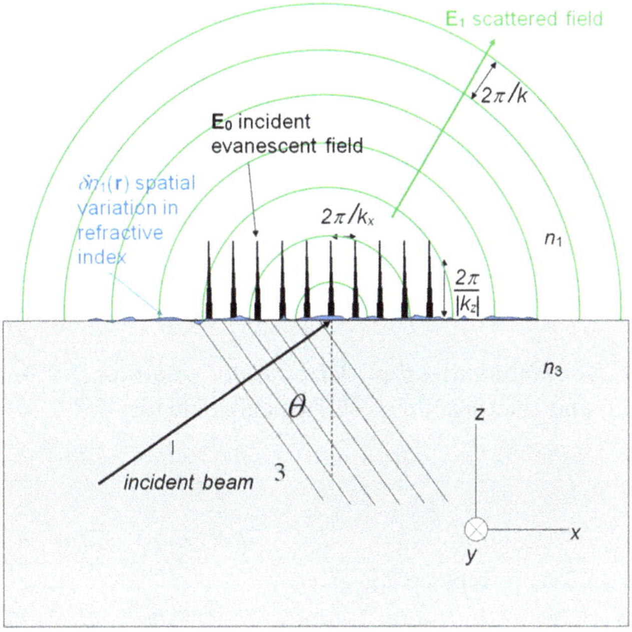

Figure 3.10. Schematic view of scattering induced by TIR, showing the key variables and the coordinate system. The infinite plane-wave incident beam is assumed to propagate in the x–z plane. The δn_1 variations refer to refractive index, not to a surface roughness as symbolically shown. The scattered light consists of both propagating and evanescent parts; the propagating part is shown (in green) with its wave spacing of $2\pi/k$. The depth of the evanescent field $2\pi/\ |k_z|$ is exaggerated here (relative to the evanescent wavefront spacing $2\pi/k_x$). For TIR, k_z is imaginary. The incident evanescent field polarization is either p-pol (along z) or s-pol (along y). Reprinted from Axelrod and Axelrod 2021, copyright (2021) with permission of Elsevier.

3.5.1 Fundamental equations

Maxwell's equations (in SI units) for a linear medium are (repeated formally from equations (1.1a–1.1d)):

$$\begin{aligned} \nabla \cdot (\varepsilon \mathbf{E}) &= \mathbf{0} \\ \nabla \cdot \mathbf{B} &= \mathbf{0} \\ \nabla \times \mathbf{E} &= -\frac{\partial \mathbf{B}}{\partial t} \\ \nabla \times \frac{\mathbf{B}}{\mu} &= \frac{\partial (\varepsilon \mathbf{E})}{\partial t}, \end{aligned} \tag{3.25a–d}$$

where (as before) the electric and magnetic fields **E** and **B** are functions of position. However (unlike before), **E** and **B** contain contributions from both an incident evanescent field and also scattered fields due to spatial variations in the refractive index. A positional dependence of the dielectric constant ε and magnetic permittivity μ accounts for those spatial variations. Thus, equations (3.25a) and (3.25d) become:

$$\varepsilon(\nabla \cdot \mathbf{E}) + \mathbf{E} \cdot \nabla\varepsilon = 0$$
$$\frac{1}{\mu}(\nabla \times \mathbf{B}) + \nabla\left(\frac{1}{\mu}\right) \times \mathbf{B} = \varepsilon\frac{\partial \mathbf{E}}{\partial t}. \tag{3.26a–b}$$

If we assume that the dielectric medium is non-magnetic, then $\mu = \mu_0$ (constant over all space), so $\nabla\left(\frac{1}{\mu}\right) = 0$ and equation (3.26b) becomes

$$(\nabla \times \mathbf{B}) = \varepsilon\mu_0\frac{\partial \mathbf{E}}{\partial t}. \tag{3.27}$$

To eliminate **B**, we combine the time derivative of equation (3.27) with the curl of equation (3.25c), and then use the vector calculus identity for $\nabla \times \nabla \times \mathbf{E}$, thereby obtaining

$$\nabla(\nabla \cdot \mathbf{E}) - \nabla^2\mathbf{E} = -\varepsilon\mu_0\frac{\partial^2 \mathbf{E}}{\partial t^2}. \tag{3.28}$$

Using equation (3.26a) to replace $\nabla\mathbf{E}$ gives:

$$\nabla\left(\mathbf{E} \cdot \frac{\nabla\varepsilon}{\varepsilon}\right) = \varepsilon\mu_0\frac{\partial^2 \mathbf{E}}{\partial t^2} - \nabla^2\mathbf{E}. \tag{3.29}$$

The refractive index (here spatially dependent) is $n_1 = c\sqrt{\varepsilon\mu_0}$, where c is the vacuum speed of light. Equation (3.29) then becomes

$$\nabla\left(\mathbf{E} \cdot \frac{\nabla(n_1^2)}{n_1^2}\right) = \frac{n_1^2}{c^2}\frac{\partial^2 \mathbf{E}}{\partial t^2} - \nabla^2\mathbf{E}. \tag{3.30}$$

Equation (3.30) can be rewritten in terms of the small spatial variations of n_1 and the consequent alteration of **E** arising from those variations. We write the refractive index as a spatial variation around some constant mean value $\langle n_1\rangle$

$$n_1(\mathbf{r}) = \langle n_1\rangle + s\delta n_1(\mathbf{r}), \tag{3.31}$$

where **r** is a position in 3D space and s is a dimensionless 'smallness' parameter used to keep track of perturbation orders and which will ultimately be set to unity. Note that each power of s can be associated with the same power in $\delta n_1/\langle n_1\rangle$. If we assume that $\delta n_1(\mathbf{r}')/\langle n_1\rangle \ll 1$ (i.e. the refractive-index variations are small), then we can expand both the numerator and denominator of $\nabla(n_1^2)/n_1^2$ of equation (3.30) into a Taylor series, suppressing terms in s^2 or higher:

$$\begin{aligned}\frac{\nabla(n_1^2)}{n_1^2} &= \frac{\nabla(\langle n_1\rangle^2 + 2s\langle n_1\rangle\delta n_1 + \ldots)}{\langle n_1\rangle^2\left(1 + 2s\frac{\delta n_1}{\langle n_1\rangle} + \ldots\right)} \approx \frac{2s\nabla(\delta n_1)}{\langle n_1\rangle}\left(1 - 2s\frac{\delta n_1}{\langle n_1\rangle}\right) \\ &\approx \frac{2s\nabla(\delta n_1)}{\langle n_1\rangle}.\end{aligned} \tag{3.32}$$

The electric field **E** given by equation (3.6) can be written as a perturbative series, whose first two terms are:

$$\mathbf{E} \cong \mathbf{E}_0 + s\mathbf{E}_1 + (\text{higher orders in } s) \tag{3.33}$$

Inserting equations (3.32) and (3.33) (to the first order in s) into equation (3.30) produces several terms including terms of the zeroth, first, and second order in s. After again suppressing terms in s^2 or higher, those terms with the same order in s can be collected into two separate but coupled equations:

$$\begin{aligned}
&(\text{zeroth order in } s) \quad \frac{\langle n_1\rangle^2}{c^2}\frac{\partial^2(\mathbf{E}_0)}{\partial t^2} - \nabla^2(\mathbf{E}_0) = 0\\
&(\text{first order in } s) \quad \frac{\langle n_1\rangle^2}{c^2}\frac{\partial^2(\mathbf{E}_1)}{\partial t^2} - \nabla^2(\mathbf{E}_1) = \\
&2\nabla\left(\mathbf{E}_0 \cdot \frac{\nabla(\delta n_1)}{\langle n_1\rangle}\right) - \frac{2\langle n_1\rangle^2}{c^2}\frac{\delta n_1}{\langle n_1\rangle}\frac{\partial^2(\mathbf{E}_0)}{\partial t^2}.
\end{aligned} \tag{3.34a and b}$$

The general solution of the homogeneous zeroth-order equation (3.34a) is the field configuration in the absence of refractive-index variations; in our application, this is the incident evanescent light in medium 1. The particular solution of the first-order non-homogeneous equation (3.34b) is the light scattered from the incident evanescent field due to the spatial refractive-index variations, provided that the particular solution contains no additive term which solves the corresponding homogeneous equation.

The first-order linear approximation that leads to equation (3.34b) predicts that if $\delta n_1/\langle n_1\rangle$ is zero everywhere, then $\mathbf{E}_1 = 0$ as expected. Equation (3.34b) is most accurate for $|\mathbf{E}_1| \ll |\mathbf{E}_0|$. In principle, scattered light can rescatter. In the $\mathbf{E}_1 << \mathbf{E}_0$ approximation implied by equation (3.34b), rescattering is implicitly approximated as negligible. The theory also implicitly assumes that scattered light $\mathbf{E}_1$ does not interact with or reflect off the TIR surface.

3.5.2 Parameter definitions

We define the time Fourier transforms of $\mathbf{E}_0$ and $\mathbf{E}_1$ as follows:

$$\tilde{\mathbf{E}}_{0,1}(\mathbf{r}, \omega) = \frac{1}{(2\pi)^{1/2}}\int \mathbf{E}_{0,1}(\mathbf{r}, t)e^{-i\omega t}dt. \tag{3.35}$$

The time Fourier transform of equation (3.34b) is now:

$$\frac{\langle n_1\rangle^2\omega^2}{c^2}\tilde{\mathbf{E}}_1 + \nabla^2\tilde{\mathbf{E}}_1 = -2\nabla\left(\tilde{\mathbf{E}}_0 \cdot \frac{\nabla(\delta n_1)}{\langle n_1\rangle}\right) - \frac{2\langle n_1\rangle^2\omega^2}{c^2}\frac{\delta n_1}{\langle n_1\rangle}\tilde{\mathbf{E}}_0. \tag{3.36}$$

The electric field's scalar wavenumber for propagating light in the medium n_1 is

$$k \equiv \frac{\omega\langle n_1\rangle}{c} = k_0\langle n_1\rangle, \tag{3.37}$$

where $k_0 = 2\pi/\lambda_0$ and λ_0 is the vacuum wavelength. For abbreviation, we define the spatially dependent variation of the refractive index as a fraction:

$$\gamma(\mathbf{r}) \equiv \frac{\delta n_1(\mathbf{r})}{\langle n_1 \rangle}. \tag{3.38}$$

Rewriting equation (3.36) in terms of these definitions, we obtain:

$$(\nabla^2 + k^2)\tilde{\mathbf{E}}_1 = -2\nabla(\tilde{\mathbf{E}}_0 \cdot \nabla\gamma) - 2k^2\gamma\tilde{\mathbf{E}}_0 \tag{3.39}$$

which is an inhomogeneous Helmholtz equation.

Although not necessary for the general theory but of important consequence for our particular results, we assume that the unscattered incident light $\mathbf{E}_0$ is evanescent and decays exponentially with decay factor $\exp(ik_z z)$ in the z direction (z is normal to the planar TIR surface which is oriented in the x–y plane), where k_z is positive and imaginary:

$$\tilde{\mathbf{E}}_0(\mathbf{r}') = E_0\hat{\mathbf{e}}_0 e^{i(k_x x' + k_y y' + k_z z')}. \tag{3.40}$$

The incident field $\mathbf{E}_0$ polarization, denoted by the unit vector $\hat{\mathbf{e}}_0$, can be either p-pol (i.e. approximated as $\mathbf{E}_{0p} = E_{0p}\hat{\mathbf{z}}$ and oriented parallel to the x–z plane of propagation) or s-pol (i.e. $\mathbf{E}_{0s} = E_{0s}\hat{\mathbf{y}}$, oriented perpendicular to the x–z plane of propagation).

The wavefronts of $\mathbf{E}_0$ are assumed to travel in the x–z plane, to be sinusoidal in the x direction, and to be constant in the y direction, so that $k_y = 0$. The wavefront spacing of $\mathbf{E}_0$ in the x direction is $2\pi/k_x$. Since we are assuming evanescence, this wavelength spacing is smaller than the propagating wavelength $2\pi/k$. The relationship between these incident electric field wavenumbers is:

$$k^2 = k_x^2 + k_z^2. \tag{3.41}$$

Even before it has been solved, equation (3.39) can be interpreted semi-qualitatively. The two terms on the right correspond to distinct physical origins of scattering. The second term on the right is proportional to γ and can be zero in principle only if γ is uniformly zero over all 3D space. The first term on the right is related to the extent to which the incident electric field polarization aligns with the local gradient of γ (or, in other words, 'cuts across lines' of constant γ). In some polarizations and geometries, this term may equal zero. To see this, briefly assume for simplicity that γ does not vary in the y direction, so the gradient component in that direction is zero. Then for $\mathbf{E}_0$ in the y direction (i.e. s-pol), that whole first term is zero.

3.5.3 Green's function solution for the perturbative approach

The Green's function for the Helmholtz operator on the left-hand side of equation (3.39) is

$$G(\mathbf{r}, \mathbf{r}') = G(\mathbf{r} - \mathbf{r}') = -\frac{e^{ik|\mathbf{r}-\mathbf{r}'|}}{4\pi\,|\,\mathbf{r} - \mathbf{r}'|}. \tag{3.42}$$

The primed positions $\mathbf{r}'$ refer to the 'source' (here, the refractive-index variations in the evanescent field near the n_1:n_3 interface). The unprimed positions $\mathbf{r}$ refer to the 'observation' point (here, anywhere in the n_1 medium). Equation (3.42) is well-defined everywhere except where the observation point is exactly at the source point $\mathbf{r} = \mathbf{r}'$ (called 'local'). Qualitatively, $G(\mathbf{r},\mathbf{r}')$ is essentially the electric field amplitude and phase observed at position $\mathbf{r}$ as generated by a point source located at $\mathbf{r}'$, so long as $r \neq r'$ (called 'nonlocal'). Quantitatively, it is the solution of the left-hand side of equation (3.15) as would be 'driven' by a delta-function spike on the right side:

$$(\nabla^2 + k^2)G(\mathbf{r} - \mathbf{r}') = \delta(\mathbf{r}'). \tag{3.43}$$

The nonlocal solution of equation (3.39) then becomes:

$$\tilde{\mathbf{E}}_1(\mathbf{r} \neq \mathbf{r}') = -2\int d^3\mathbf{r}' \;\; G(\mathbf{r} - \mathbf{r}')\left\{\nabla'\left[\tilde{\mathbf{E}}_0(\mathbf{r}') \cdot \nabla'\gamma(\mathbf{r}')\right] + k^2\gamma(\mathbf{r}')\tilde{\mathbf{E}}_0(\mathbf{r}')\right\}. \tag{3.44}$$

Equation (3.44) shows that the first-order correction due to scattering arises from two causes: the gradient of the refractive index in the sample, and the deviation of the refractive index at any point from a global average. The gradient term (the first in the integral's brackets) is relatively important for samples containing refractive-index variations of higher spatial frequencies; the non-gradient term is relatively more important for low spatial frequencies.

Given a known refractive-index pattern $\gamma(\mathbf{r}')$, equation (3.44) could be used directly (presumably by digital integration) to calculate $\mathbf{E}_1$. However, it is not the complete solution because the integration is valid only for $\mathbf{r} \neq \mathbf{r}'$. The goal of the next two subsections is to transform equation (3.44) into a different form, one to which the solution for $\mathbf{r} = \mathbf{r}'$ can be added easily. An alternative approach that is more mathematically rigorous, involving integration by parts, is presented in Axelrod and Axelrod (2021).

3.5.4 Inclusion of the local case $\mathbf{r} = \mathbf{r}'$

The refractive-index spatial deviation pattern γ $(\mathbf{r}')$ can be written as a series of 3D delta functions, infinitesimally closely spaced:

$$\gamma(\mathbf{r}') = \int d^3\mathbf{r}''\delta(\mathbf{r}' - \mathbf{r}'')\gamma(\mathbf{r}''). \tag{3.45}$$

Inserting equation (3.45) into equation (3.44) gives:

$$\tilde{\mathbf{E}}_1(\mathbf{r} \neq \mathbf{r}') = -2\int d^3\mathbf{r}''\gamma(\mathbf{r}'')\int d^3\mathbf{r}' G(\mathbf{r} - \mathbf{r}')\left\{\nabla'\left[\tilde{\mathbf{E}}_0(\mathbf{r}') \cdot \nabla'\delta(\mathbf{r}' - \mathbf{r}'')\right] + k^2\delta(\mathbf{r}' - \mathbf{r}'')\tilde{\mathbf{E}}_0(\mathbf{r}')\right\}. \tag{3.46}$$

The integral over $d\mathbf{r}'$ can be performed explicitly. The result, for each component i of the vector $\tilde{\mathbf{E}}_1(\mathbf{r} \neq \mathbf{r}')$, is:

$$\begin{aligned}\widetilde{E}_{1,i}(\mathbf{R} \neq 0) = -2\int d^3\mathbf{r}''\gamma(\mathbf{r}'')\widetilde{E}_{0,i}(\mathbf{r}'')\frac{e^{ikR}}{4\pi R^3}\\ \sum_{j=1}^{3}\left[-\delta_{ij}(1 - ikR - k^2R^2) + (3\text{–}3ikR - k^2R^2)\frac{R_iR_j}{R^2}\right],\end{aligned} \tag{3.47}$$

where:

(a) $\mathbf{R} = \mathbf{r} - \mathbf{r}'$ and $R = |\mathbf{r} - \mathbf{r}'|$;
(b) the subscripts i and j (= 1,2,3) refer to the Cartesian components (x,y,z, respectively) of $\mathbf{R}$ and $\mathbf{E}$;
(c) the incident field $\mathbf{E}_0$ polarization can be either p-pol (i.e. $\mathbf{E}_{0p} = E_{0,1}\hat{\mathbf{x}} + E_{0,3}\hat{\mathbf{z}}$, oriented parallel to the x–z plane of propagation) or s-pol (i.e. $\mathbf{E}_{0s} = E_{0,2}\hat{\mathbf{y}}$, oriented perpendicular to the x–z plane of propagation);
(d) δ_{ij} is a Kronecker delta (= 1 if subscripts $i = j$, and = 0 if $i \neq j$);
(e) $i = (-1)^{1/2}$ where it is not a subscript.

Note that the (nonlocal) electric field $\mathbf{\Phi}$ around a small oscillating dipole vector $\hat{\mathbf{p}}$ of unit amplitude (i.e. $\hat{\mathbf{p}} \equiv \mathbf{p}/|\mathbf{p}|$) embedded in a dielectric medium is (from standard electricity and magnetism texts):

$$\mathbf{\Phi}(\mathbf{R} \neq 0) = \frac{1}{4\pi\varepsilon}\left\{k^2\left(\hat{\mathbf{R}} \times \hat{\mathbf{p}}\right) \times \hat{\mathbf{R}}\frac{e^{ikR}}{R} + \left[3\hat{\mathbf{R}}\left(\hat{\mathbf{R}} \cdot \hat{\mathbf{p}}\right) - \hat{\mathbf{p}}\right]\left(\frac{1}{R^3} - \frac{ik}{R^2}\right)e^{ikR}\right\}, \quad (3.48)$$

where $\varepsilon = \varepsilon(\mathbf{r}')$is the permittivity of the space in which the dipole resides. If $\hat{\mathbf{p}} \equiv \hat{i}$, where i =1,2,3 represents unit vectors in the x,y, and z directions, then the expansion of this dipole vector field into its components produces the exact same $\sum_j$ factor as does equation (3.47):

$$\Phi_i(\mathbf{R} \neq 0) = \frac{e^{ikR}}{4\pi\varepsilon R^3}\sum_{j=1}^{3}\left[-\delta_{ij}(1 - ikR - k^2R^2) + (3 - 3ikR - k^2R^2)\frac{R_iR_j}{R^2}\right]. \quad (3.49)$$

Substituting this expression for an oscillating dipole into the expression for first-order scattering in equation (3.47) gives

$$\widetilde{E}_{1,i}(\mathbf{R} \neq 0) = -2\int d^3\mathbf{r}''\gamma(\mathbf{r}'')\widetilde{E}_{\mathbf{0},i}(\mathbf{r}'')\varepsilon\Phi_i(\mathbf{R} \neq 0). \quad (3.50)$$

Equation (3.50) has an intuitive physical interpretation. It says that the first-order scattered field can be expressed as an integral over single oscillating dipole electric fields that are induced by $\mathbf{E}_0$, with a weighting in accordance with the local deviation of the refractive index.

But most importantly, $\mathbf{E}_1$ is now in a form that allows us to take into account the local (R=0) contribution to the first-order scattering. According to Jackson (1999), the full field expression for a unit vector static dipole (corresponding to k=0 in equation (3.48) here) requires an additional δ-function term for the extreme near field:

$$\mathbf{\Phi}_{k=0}(all\mathbf{R}) = \frac{1}{4\pi\varepsilon}\left[\frac{3\hat{\mathbf{R}}\left(\hat{\mathbf{R}} \cdot \hat{\mathbf{p}}\right) - \hat{\mathbf{p}}}{R^3}\right]_{R\neq 0} - \frac{\hat{\mathbf{p}}\delta(\mathbf{R})}{3\varepsilon}. \quad (3.51)$$

We now assume that this additional term is also valid for any $k > 0$, since the field of an oscillating dipole in the extreme near field is identical to that of a static field except for the time variation $\exp(-i\omega t)$:

$$\mathbf{\Phi}(all\mathbf{R}) = \frac{1}{4\pi\varepsilon}\left\{k^2(\hat{\mathbf{R}} \times \hat{\mathbf{p}}) \times \hat{\mathbf{R}}\frac{e^{ikR}}{R} + \left[3\hat{\mathbf{R}}(\hat{\mathbf{R}} \cdot \hat{\mathbf{p}}) - \hat{\mathbf{p}}\right]\left(\frac{1}{R^3} - \frac{ik}{R^2}\right)e^{ikR}\right\}_{R\neq 0} - \frac{\hat{\mathbf{p}}\delta(\mathbf{R})}{3\varepsilon}. \quad (3.52)$$

This extra local term can then be used to augment the nonlocal expression for E_1 (equation 3.50) to complete its validity for all zones:

$$\begin{aligned}\widetilde{E}_{1,\,i}(all\mathbf{R}) = -2 \int d^3\mathbf{r}''\gamma(\mathbf{r}'')\widetilde{E}_{\mathbf{0},\,i}(\mathbf{r}'')\\ \varepsilon[\Phi_i(\mathbf{R} \neq 0)] - 2 \int d^3\mathbf{r}''\gamma(\mathbf{r}'')\widetilde{E}_{\mathbf{0},\,i}(\mathbf{r}'')\left[-\frac{\hat{\mathbf{p}}\delta(\mathbf{R})}{3}\right].\end{aligned} \quad (3.53)$$

Expressed in the formalism of equations (3.47) and (3.49), this becomes

$$\begin{aligned}\widetilde{E}_{1,\,i}(\mathbf{r}) = 2 \int d^3\mathbf{r}''\gamma(\mathbf{r}'')\widetilde{E}_{\mathbf{0},\,i}(\mathbf{r}'')\frac{e^{ikR}}{4\pi R^3}\Bigg|_{R\neq 0}\\ \sum_{j=1}^{3}\left[-\delta_{ij}\left(1 - ikR - k^2R^2\right) + \left(3 - 3ikR - k^2R^2\right)\frac{R_iR_j}{R^2}\right]_{R\neq 0} - \frac{2\gamma(\mathbf{r})\widetilde{E}_{\mathbf{0},\,i}(\mathbf{r})}{3}.\end{aligned} \quad (3.54)$$

Equation (3.54) is the main theoretical result: it can be used to generate $\mathbf{E}_1$ from a variety of input $\gamma(\mathbf{r}')$ refractive-index patterns, both hypothetical and experimentally measured. Equation (3.54) has two main terms. The first, an integral over source positions $\mathbf{r}'$, represents a *nonlocal* effect: a refractive-index disturbance at one location affects the scattered field observable at another location. The $\exp(ikR)$ factor, when combined with the $\exp(-i\omega t)$ that was suppressed by the time Fourier transform of equation (2.36), produces outward traveling propagating waves. Some of these waves may propagate far enough to reduce surface selectivity in TIRF. However, other such waves have R^{-m} dependencies where $m>1$; these waves decay rather quickly with increasing R and tend to preserve surface selectivity. The second term in equation (3.54) represents a *local* effect, present only at the source disturbance. Moreover, it does not propagate: if the incident $\mathbf{E}_0(\mathbf{r})$ is evanescent in the z direction, this second term is also evanescent. It only increases or decreases the amplitude of the incident evanescent field, depending on relative phase and sign. Nonetheless, both terms are important in quantitatively reporting surface selectivity.

3.5.5 Reporting surface selectivity: intensity and evanescent depth

For the efficient excitation of fluorophores (e.g. in TIRF microscopy), the relevant quantity is the total intensity I_{tot}:

$$I_{\text{tot}}(\mathbf{r}) = |\mathbf{E}_0 + \mathbf{E}_1|^2. \quad (3.55)$$

I_{tot} accounts for interference effects between the incident and scattered light, since their phases are generally different at any observation location $\mathbf{r}$.

If there is no scattering at all ($\mathbf{E}_1 = 0$), then the intensity is uniform over x–y and exponentially decaying in z, with an e^{-2} depth denoted by ρ_0:

$$I_0(\mathbf{r}) = |\mathbf{E}_0|^2 = I_0 e^{-2z/\rho_0}. \quad (3.56)$$

For an interface between a lower-refractive-index material n_1 and a TIR substrate material n_3, and for incidence angle θ_3 and vacuum wavelength λ_0, the depth ρ_0 is:

$$\rho_0 = \frac{\lambda_0}{2\pi n_3\sqrt{\sin^2\theta_3 - \left(\frac{n_1}{n_3}\right)^2}}. \tag{3.57}$$

The more standard measure of pure evanescent depth (equation (2.12)), usually denoted as d, is at the e^{-1} point, which is exactly half of ρ_0, the e^{-2} point. The e^{-2} point is more suitable for measuring surface selectivity: for the zero-scattering case and a uniform distribution of fluorophores in z, 86% of TIR fluorescence originates from within ρ_0 but only 63% from within d.

In the presence of scattering, I_{tot} is not necessarily exponentially decaying in z. However, a good measure of the effect of scattering is the z distance where I_{tot} drops to less than e^{-2} of its value at z=0, and how that depth $\boldsymbol{\rho}$ compares with the scattering-free $\boldsymbol{\rho}_0$. 'Surface selectivity' can be defined as $\rho(x,y)/\rho_0$, where

$$\frac{I_{total}(x, y, \rho)}{I_{total}(x, y, 0)} = e^{-2}. \tag{3.58}$$

A calculation of surface selectivity is displayed here (figure 3.11) for two different samples: a hypothetical cylinder that has a radius of 2000 nm oriented along the z-axis (i.e. perpendicular to the TIR surface) and a cell culture whose refractive index was experimentally measured using the method of Bohannon *et al* (2017).

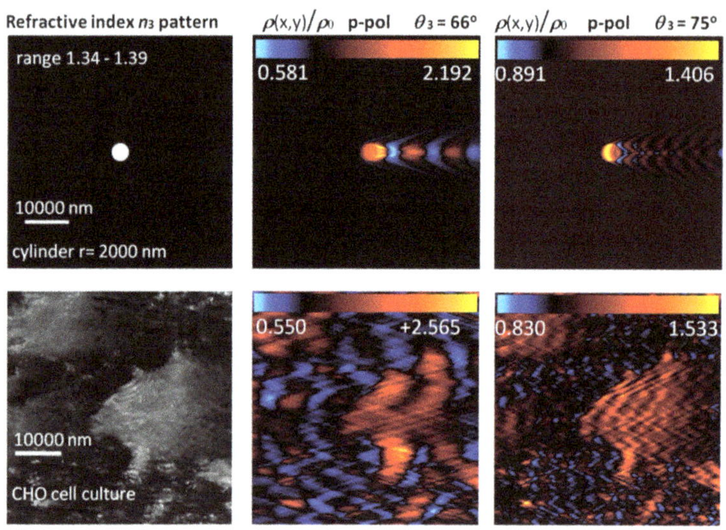

Figure 3.11. Surface selectivity ratio $\rho(x,y)/\rho_0$, calculated using equations (3.54), (3.55), and (3.58). In both the cylinder (top row) and the Chinese hamster ovary cell culture (bottom row), the input range of refractive indices varies from a low of 1.34 to a high of 1.39. The incident evanescent field travels from left to right, generated by a TIR beam at incidence angle $\boldsymbol{\theta}_3$. The scale in any color bar is linear (i.e. a color halfway through the bar represents a value halfway between the indicated minimum and maximum for $\rho(x,y)/\rho_0$). The colors are coded red-to-yellow for $\rho(x,y)/\rho_0 > 1$, blue-to-aqua for $\rho(x,y)/\rho_0 < 1$, and black for $\rho(x,y)/\rho_0 = 1$.

It is clear that higher incidence angles produce less scattering. For example, on cells at barely supercritical incidence angles, the maximum e^{-2} depth is 2.565 times what it would be with no scattering; at higher angles, the e^{-2} depth is only 1.533 times what it would be with no scattering. Both cases retain significant surface selectivity, but higher incidence angles are better.

Surface selectivity is particularly relevant to the 'z resolution' of the TIRF microscopy technique (covered starting in the next chapter). TIRF microscopy offers an improvement in z resolution of about a factor of five, compared to confocal or multiphoton microscopies. Therefore, if $\rho(x,y)/\rho_0$ is, say, less than three, a significant advantage in z resolution is still preserved by TIRF. (Of course, the performance of confocal and multiphoton microscopies is also decremented by scattering.)

As an approximate alternative to the above presentation, the scattering could be presented as a constant (z-independent term with weight α_{scatt}) added to the pure evanescent exponential decay expected without scattering (i.e. with its $1/e^2$ point at $z = \rho_0$) with a weighting of $(1-\alpha_{\text{scatt}})$. The goal is to find a value of α_{scatt} consistent with the surface selectivity ratio ρ/ρ_0 reported in figure 3.11:

$$(1 - \alpha_{\text{scatt}})e^{-2(\rho/\rho_0)} + \alpha_{\text{scatt}} = e^{-2}. \tag{3.59}$$

Therefore,

$$\alpha_{\text{scatt}} = \frac{e^{-2} - e^{-2(\rho/\rho_0)}}{1 + e^{-2(\rho/\rho_0)}}. \tag{3.60}$$

For $\rho/\rho_0 = 2$ (the typical maximum typical for the above cells at intermediate supercritical incidence angles), $\alpha_{\text{scatt}} = 0.12$. In other words, a rough approximation of the effect of scattering is that it adds a constant amount to the evanescent decay with a relative maximum amplitude of about 12%. Since that is a maximum (where the scattering is the worst), a reasonable guess for other areas is 10%.

Parts of this chapter have appeared in Axelrod D 2012 Fluorescence excitation and imaging of single molecules near dielectric-coated and bare surfaces: a theoretical study *J. Microsc.* **247** 147–60.

Further reading

Axelrod D 2012 Fluorescence excitation and imaging of single molecules near dielectric-coated and bare surfaces: a theoretical study *J. Microsc.* **247** 147–60

Axelrod J and Axelrod D 2021 Light scattering in TIRF microscopy: a theoretical study of the limits to surface selectivity *Biophys. J.* **120** 2952–68

Bohannon K P, Holz R W and Axelrod D 2017 Refractive index imaging of cells with variable-angle near TIR microscop *Microsc. Microanal.* **23** 978–88

Born E and Wolf E 1999 *Principles of Optics: Electromagnetic Theory of Propagation, Interference and Diffraction of Light* 7th edn (Cambridge: Cambridge University Press) ch XIII. This chapter, not appearing in earlier editions, is entitled 'The theory of scattering of light by inhomogeneous media'.

Jackson J D 1999 *Classical Electrodynamics* 3rd edn (New York: Wiley) section 4.1

IOP Publishing

Total Internal Reflection Fluorescence (TIRF) and Evanescence Microscopies

Daniel Axelrod

Chapter 4

Emission of fluorophores near a surface

This chapter is coauthored with Dr Edward H Hellen, University of North Carolina, Greensboro, Department of Physics.

The previous chapters concentrated on the TIR evanescent field itself, and how features in the lower-index material—such as refractive-index variations, layers, or absorption—affect the field. By far the most common use of an evanescent field in biophysics is to detect the presence of luminescent molecules (usually fluorophores) in the field. From the point of view of a luminescent molecule, the temporally sinusoidal variation of evanescent wavefronts passing any location in the field acts as a perturbation which can cause a quantum-mechanical transition from the ground state to the excited state. The luminophore does not 'know' that the field is evanescent; it only knows the local strength of the field and its frequency. The probability of such an excitatory transition is, to the first order, linearly proportional to local intensity $I(z)$ of the field. A (generally rapid but random) return to the ground state is then accompanied by the release of a photon, usually of a lower-frequency color because of the loss of some energy to molecular vibrations. So, in general, the total amount of emitted fluorescence power $\mathcal{F}$ is proportional to the local evanescent intensity I:

$$\mathcal{F}(z) = \alpha(z)I(z). \tag{4.1}$$

This simple expression conceals a whole slew of complications.

(a) $I(z)$ may be a function of x and y as well as z. For example, if two coherent evanescent fields arise from intersecting beams at the TIR interface, a periodic striped x–y interference pattern is created in I. This feature will be discussed in the sections on 'structured illumination' microscopy (SIM) as applied to TIR in chapters 5 and 6.

doi:10.1088/978-0-7503-3351-1ch4

(b) The emission power $\mathcal{F}$ does not occur isotropically: the emission intensity is brighter in some directions than in others. The emission light actually 'seen' by a detector that gathers light from a particular range of directions is a nontrivial outcome of detector size and location. In addition, the detected polarization is related to, but not generally the same as, the polarization of I. The part of $\mathcal{F}$ that is actually detected is called the 'fluorescence' F.

(c) The proportionality factor α is a complicated function of proximity to the TIR surface and also the intrinsic quantum efficiency of the fluorophore, which may itself be a function of its chemical environment, including proximity to the surface.

In this chapter, we consider the properties of the emission, given that the fluorophore has already been excited. The fluorescence F is then simply proportional to a product of the probability of excitation and the probability of detection.

A note on notation: where emission electric field properties are discussed in general, no subscript is used. Where we refer to an electric field that is specific to a particular medium, for example, medium 1 or medium 3, subscripts are used.

4.1 The emission near field: a semi-qualitative view

A single fluorophore that is repeatedly excited and emitting can be modeled as a classical oscillating dipole, and the emission (and its consequent evanescent aspects) can be understood using classical electromagnetism theory. In chapter 1, we discussed how 'squeezing' classical wavefronts to a spacing closer than would be expected in freely propagating light leads to evanescence. Apart from evanescence in excitation light, light emitted from a fluorophore can also show observable effects of evanescence. Even if there are no interfaces nearby, the spacing between wavefronts is squeezed if the size of the light source is very small, i.e. subwavelength. The result is that a portion of the light field is evanescent in the direction away from the source. This portion is called the 'near field' and it has a number of practical consequences and analogies to excitation evanescence.

To see the necessity of including wavelengths that are shorter than the wavelength of the corresponding propagating light at that frequency, consider an emission point source of EM radiation, an oscillating dipole. The electric field at a distance r from an oscillating dipole vector $\boldsymbol{\mu}$ located at the origin in a homogeneous medium with refractive index n_1 is directly derivable from Maxwell's equations. If the $\exp(-i\omega t)$ time factor is suppressed, it can be written in exact form (i.e. not an approximation) as follows:

$$\breve{\mathbf{E}}(\mathbf{r}) = \frac{e^{ik_1 r}}{4\pi\varepsilon}k^3\left[(\hat{\mathbf{r}}\times\boldsymbol{\mu})\times\hat{\mathbf{r}}\left(\frac{1}{kr}\right) - (3\hat{\mathbf{r}}(\hat{\mathbf{r}}\cdot\boldsymbol{\mu}) - \boldsymbol{\mu})\left(\frac{i}{(kr)^2}\right)\right.$$
$$\left. + (3\hat{\mathbf{r}}(\hat{\mathbf{r}}\cdot\boldsymbol{\mu}) - \boldsymbol{\mu})\left(\frac{1}{(kr)^3}\right)\right] - \frac{\boldsymbol{\mu}\delta(r)}{3\varepsilon}, \tag{4.2}$$

where the wavenumber k is as defined in equation (1.8), except here it refers to the wavenumber of the emission light rather than the excitation light, and where ε is the permittivity of the medium in which the dipole is embedded. This is essentially the same as equation (3.52), except that there, $\boldsymbol{\Phi}$ referred to the scattering field produced by an induced dipole **p**. Here, we refer to the field $\breve{\mathbf{E}}$ generated by an emission dipole **μ**. Concerning the notation, the upward-curving 'hat' on the $\breve{\mathbf{E}}$ refers to fluorescence emission, to distinguish it from the hatless **E** used in previous chapters for non-fluorescent incident, reflected, or scattered light.

$\breve{\mathbf{E}}(\mathbf{r})$ is an outgoing spherical wave oscillating at angular frequency ω, but whose amplitude and polarization depend upon the direction and distance to the observation point at **r**. The three terms in the square brackets have increasingly rapid rates of 'drop-off' with the distance $k_1 r$ to the observation point (measured in terms of the number of wavelengths × 2π). The first term, with the factor $(k_1 r)^{-1}$, is the standard 'far zone' light that retains the same integrated energy through any sphere centered on the fluorophore and is responsible for all of the propagating emission seen from an isolated fluorophore far (i.e. $k_1 r >> 1$) from any interfaces. The next two terms are called the 'near zone' because they decay more rapidly with distance. Note that the leading term is proportional to k^2/ε. Since both k^2 and ε are proportional to n^2 (see equation 1.6), the amplitude of that leading term has no dependence on n; it is the same, regardless of the homogeneous medium in which the dipole is embedded.

The question is: where are the 'evanescent' terms in equation (4.2)? This question becomes relevant when considering how the 'spherical' wave of equation (4.2) interacts with interfaces. Snell's law for refraction, the reflection law, and Fresnel coefficients are derived by assuming the incident light is a plane wave, not a spherical wave. Therefore, we express $\breve{\mathbf{E}}(\mathbf{r})$ as an integral over plane waves with some weighting factors. Each component plane wave interacts individually with the interface in a manner that is accurately predicted by plane-wave optics laws, and then we integrate the resulting reflected and transmitted plane waves with the original weighting factors to derive the behavior of $\breve{\mathbf{E}}(\mathbf{r})$ as it interacts with the optical system.

The resolution of a curved wavefront field into component plane waves is called the 'angular spectrum of plane waves.' This procedure is the theoretical justification for drawing 'rays' from a point source and then treating each ray's path through an optical system as if it were a plane wave. The angular spectrum of plane waves is a special case of expressing the original field as a 3D Fourier integral:

$$\breve{\mathbf{E}}(\mathbf{r}) = \oint \tilde{\breve{\mathbf{E}}}(\mathbf{k}_1)\, e^{i\mathbf{k}_1\cdot\mathbf{r}} d^3\mathbf{k}_1, \tag{4.3}$$

where the exponential term represents plane waves and the vector direction and amplitude of each plane-wave component of the complete emission field are given by the $\tilde{\breve{\mathbf{E}}}(\mathbf{k}_1)$ factor. In principle, the integration must be taken over *all* possible amplitudes and directions of wavevectors $\mathbf{k}_1$. However, recall (from chapter 1) that not all mathematically possible $\mathbf{k}_1$ vectors can be used in optics: only those for which the amplitude $|\mathbf{k}_1|$ $(= k_1 = 2\pi n_1/\lambda_0)$ is a fixed constant can be used. However, there is some flexibility: k_1^2 can be written as the sum of the squares of the components $k_{1x,y,z}$. Among the optically possible $\mathbf{k}_1$ vectors, some may have (for example) $(k_{1x}^2 + k_{1y}^2) > + k_1^2$, leading to an imaginary k_{1z} and evanescence in the z direction, which is analogous to the situation in TIRF, except that the decay starts at the z position of the fluorophore.

To intuitively see that evanescent plane waves must be included in the mix, let us see what happens if only propagating plane waves are involved. Imagine that each of the 'rays' emanating from a point is an infinite plane wave and that its outward direction is indicated by the ray. So the question is whether these 'squeezed' plane-wave components with $(k_{1x}^2 + k_{1y}^2) > k_1^2$ are really necessary to reconstruct the original field. The answer is 'yes.' Let us assume the opposite, noting that Maxwell's equations are reversible in time. We can then construct a hypothetical field consisting of propagating plane waves, not diverging away from a point, but converging from all directions toward a point where they are all in phase. This can be written as an integral over plane waves $e^{i\mathbf{k}_1\mathbf{r}}$:

$$\mathbf{E}_{\text{convg plane}}(\mathbf{r}) = E_0 \oint e^{i\mathbf{k}_1\cdot\mathbf{r}} d^3\mathbf{k}_1. \tag{4.4}$$

Unlike equation (4.3), we try to omit any squeezed plane-wave components from the integration, i.e. hypothesize that all these converging waves from all directions have equal amplitudes $k_1 = |\mathbf{k}_1|$, they all have phase zero at $\mathbf{r} = 0$, and additionally, the x, y, z components of $\mathbf{k}_1$ are all real. The integral is really over directions only, and not also over all possible amplitudes k_1, unlike the similar-looking integral expression for a Dirac delta function $\delta(\mathbf{r})$. The integral can be written explicitly:

$$\oint_{\text{angles}} e^{i\mathbf{k}_1\mathbf{r}} d^2\mathbf{k}_1 = \int_{\phi=0}^{2\pi}\int_{\theta=0}^{\pi} e^{ik_1 r\cos\theta} k_1^2 \sin\theta \, d\theta d\phi = 4\pi k_1^2 \frac{\sin(k_1 r)}{(k_1 r)}. \tag{4.5}$$

Clearly, propagating plane waves converging in phase toward some $r = 0$ do *not* add up to a sharp point, but rather to a finite pattern of maximum amplitude $4\pi k_1^2$ and width $r = \pi/k_1$ from the central maximum to the first zero. Therefore, a sharp point's radiation pattern *must* include squeezed wavefronts that are evanescent in at least one direction. This is the origin of the evanescent 'near field' near a fluorophore. The 'near *zone*' terms of equation (4.2) are not the *sole* source of the 'near *field*.' Rather, all three terms, including the 'far zone' term are responsible for setting up the near field, because all three have infinite singularities at $r = 0$.

The near-field energy surrounding an emitting fluorophore cannot be detected at a distance of more than a few wavelengths away if the fluorophore is isolated. However, the near field can interact with a nearby surface that has a higher refractive index (such as a glass coverslip). In this case, the near field can be partially converted into observable propagating light in the higher-index medium.

Figure 4.1 shows typical propagating emission plane waves described by equation (4.3) (i.e. those for which $k_{1x,y,z}$ are all real); these are represented as white wavefronts. They can propagate out into medium 1, skim along the surface, or refract into medium 3. To distinguish them, a typical evanescent component (i.e. k_{1z} imaginary) is represented in orange. If the fluorophore in medium 1 is close to an interface with a higher refractive index n_3, the 'tails' of the evanescent component intersect the interface. Those tails can create propagating light in the higher-index material if their wavefront spacing matches the spacing of the evanescent tails. Given that light propagating in medium 3 has a fixed wavelength, matching can occur only if the propagation direction is at a particular angle with respect to the normal. This angle corresponds to the supercritical incidence angle that hypothetical

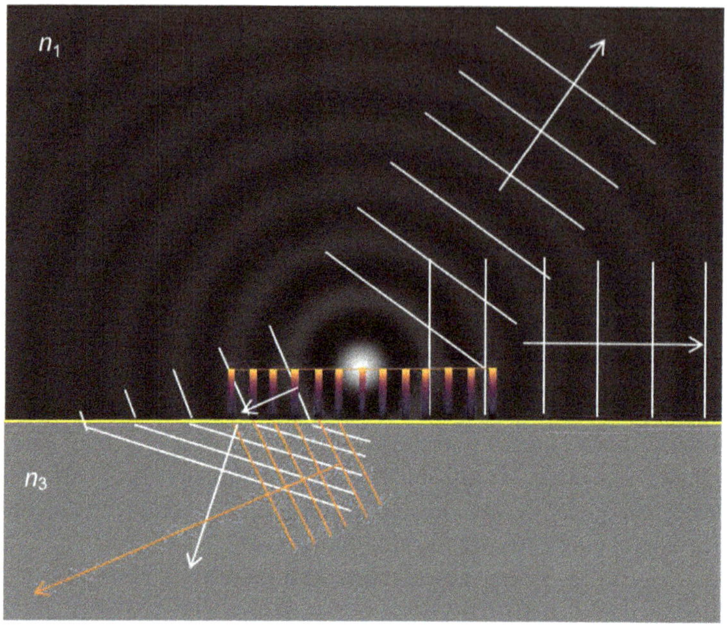

Figure 4.1. Schematic drawing of wavefronts emitted from a dipole near a surface, both propagating (white) and near field evanescent (orange). The evanescent waves in medium 1 are converted into propagating waves in medium 3 and propagate at supercritical angles.

TIR illumination would trace 'going in reverse' and setting up an evanescent field with the same squeezed wavefront spacing.

The smaller the x–y wavefront spacing of an evanescent emission component, the larger the supercritical angle of propagation in the higher-index material. But for a very small x–y spacing in medium 1, it is not possible for a propagating beam to have matching spacing at the interface. Such a field remains evanescent, even in the higher-index material, and is not normally detected. The higher the refractive index n_3, the larger the number of evanescent components of the fluorophore near field that can be captured, converted into propagating light, and detected.

4.2 Capture of the near field: summary of quantitative theory

As in section 2.2, consider two semi-infinite regions, 1 and 3, separated by an intermediate layer, region 2, of thickness h. We want to calculate the electric field vector that would be observed at any position in regions 1 or 3. The dielectric permittivities of these three regions are $\varepsilon_{1,2,3}$. The respective refractive indices are thereby $n_i = \sqrt{\varepsilon_i/\varepsilon_0}$, where ε_0 is the electric permittivity of vacuum in the SI system. We will describe the emitted field $\breve{\mathbf{E}}(\mathbf{r})$ of a fluorophore (modeled as a classical isolating dipole) embedded in the lower-index n_1 at a distance z' from the interface between regions 1 and 2. We define a 'lab' coordinate system, in which the observation point $\mathbf{r}$ is located at spherical coordinates (r,θ,ϕ) within rectangular axes $\hat{\mathbf{x}}$, $\hat{\mathbf{y}}$, $\hat{\mathbf{z}}$. The positive direction of the z-axis heads into medium 1, $z = 0$ is the boundary between regions 1 and 2, and the positive direction of the x-axis (although arbitrary) can be chosen to coincide with the direction in which TIR evanescent waves travel along the surface if TIR excitation is used.

Let each observation point **r**, together with the z-axis, define a 'plane of observation.' We now define a new rectangular coordinate system ($\hat{\mathbf{w}}$, $\hat{\mathbf{s}}$, $\hat{\mathbf{z}}$), which shares the same $\hat{\mathbf{z}}$-axis as the observation system ($\hat{\mathbf{x}}$, $\hat{\mathbf{y}}$, $\hat{\mathbf{z}}$). Axis $\hat{\mathbf{w}}$ lies within the plane of observation, and parallel to the **x–y** interface, and axis $\hat{\mathbf{s}}$ lies perpendicular to the plane of observation and parallel to the interface. Note that the orientation of the ($\hat{\mathbf{w}}$, $\hat{\mathbf{s}}$, $\hat{\mathbf{z}}$) system is just the ($\hat{\mathbf{x}}$, $\hat{\mathbf{y}}$, $\hat{\mathbf{z}}$) system rotated clockwise around $\hat{\mathbf{z}}$ by the azimuthal angle ϕ of the observation point. Figure 4.2 shows a pseudo-3D representation of these coordinate systems.

The dipole resides at a distance of z' on the positive $\hat{\mathbf{z}}$-axis. Its orientations in polar and azimuthal spherical coordinates, as measured in the ($\hat{\mathbf{x}}$, $\hat{\mathbf{y}}$, $\hat{\mathbf{z}}$) system, are (θ',ϕ'). In the ($\hat{\mathbf{w}}$, $\hat{\mathbf{s}}$, $\hat{\mathbf{z}}$) system, the dipole $\boldsymbol{\mu}$ can be written as:

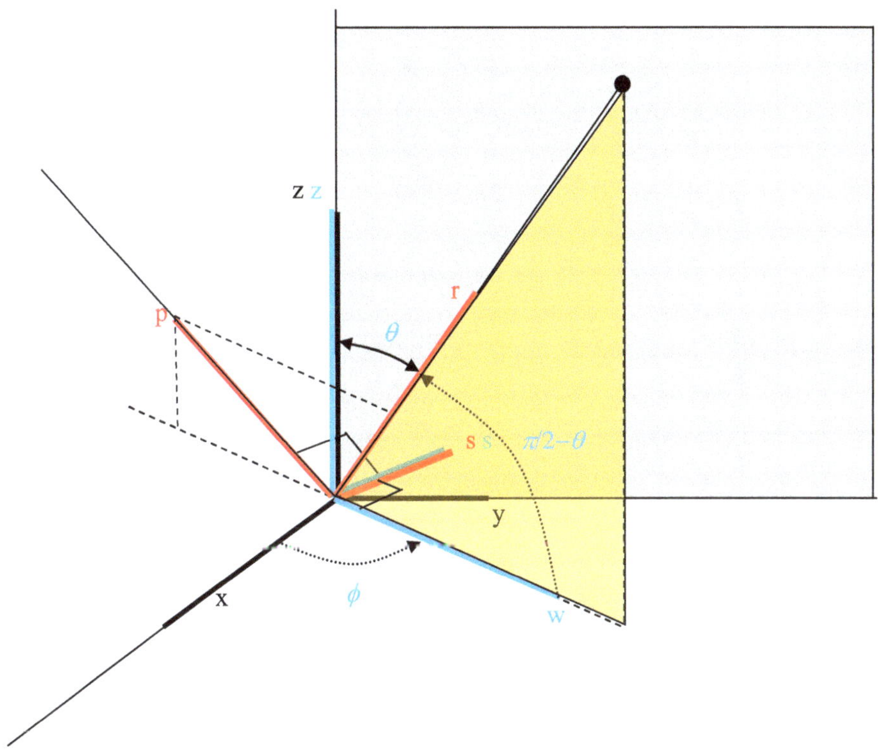

Figure 4.2. The ($\hat{\mathbf{x}}$, $\hat{\mathbf{y}}$, $\hat{\mathbf{z}}$) lab frame is shown in black. The n_1:n_3 interface is the x–y plane. The dipole resides at, or very near above, the origin. The observation point at (r,θ,ϕ) is shown as a black dot (which could be either in medium n_1, where $0 < \theta < \pi/2$ as shown, or in medium n_3, where $\pi/2 < \theta < \pi$). The ($\hat{\mathbf{w}}$, $\hat{\mathbf{s}}$, $\hat{\mathbf{z}}$) frame is shown in blue. It is obtained by rotation around **z** by an azimuthal angle of ϕ, so that the plane of observation (shown in yellow) contains the z-axis and the observation point. The ($\hat{\mathbf{r}}$, $\hat{\mathbf{s}}$, $\hat{\mathbf{p}}$) frame (useful in the following section which discusses polarization) is shown in red. It is obtained by a rotation of the ($\hat{\mathbf{w}}$, $\hat{\mathbf{s}}$, $\hat{\mathbf{z}}$) frame around **s** by an angle ($\pi/2$ − θ).

$$\boldsymbol{\mu} = \mu_w \hat{\mathbf{w}} + \mu_s \hat{\mathbf{s}} + \mu_z \hat{\mathbf{z}}$$

where

$$\begin{aligned} \mu_w &= \mu \sin\theta' \cos(\phi' - \phi) \\ \mu_s &= \mu \sin\theta' \sin(\phi' - \phi) \\ \mu_z &= \mu \cos\theta'. \end{aligned} \tag{4.6}$$

To calculate $\breve{\mathbf{E}}(\mathbf{r})$ for a dipole near a surface, we express $\breve{\mathbf{E}}(\mathbf{r})$ in terms of its plane-wave components $\breve{\breve{\mathbf{E}}}(\mathbf{k}_1)$ (equation 4.3), some of which are propagating and some of which are evanescent. Some of those propagating plane waves head directly out into medium 1 without interacting with the surface. Some of the other plane-wave components interact with the surface. For any particular observation point in medium 1, two plane waves head exactly in the (θ,ϕ) direction: a direct wave from the dipole, and a reflected wave from the surface (a combination of reflections from the 1–2 and 2–3 interfaces). For any particular observation point in medium 3, one plane wave heads exactly in the (θ,ϕ) direction (after undergoing refraction through the interface).

The way in which (or whether) a particular plane wave emitted from the dipole interacts with the surface depends on its incidence angle $\beta = \pi - \theta$ with respect to the surface normal in medium 1. For any emitted plane wave with wave vector $\mathbf{k}_1$,

$$\mathbf{k}_1 = k_x \hat{\mathbf{x}} + k_y \hat{\mathbf{y}} + k_z \hat{\mathbf{z}} \tag{4.7}$$

and

$$k_1 \equiv |\mathbf{k}_1| = \left(k_x^2 + k_y^2 + k_z^2\right)^{1/2}. \tag{4.8}$$

where, as previously, we omit the k components' subscript 1 where there is no ambiguity. Then

$$\sin\beta = \frac{k_x^2 + k_y^2}{k_1} \equiv v, \tag{4.9}$$

where v is an abbreviation to be used later. (Note that $\sin\beta = v$ can be greater than one if k_z is imaginary (and thereby $k_z^2 < 0$), which is the case for evanescent plane waves emitted from the dipole. In mathematical formalism, this means that the angle β is imaginary for evanescent plane waves.)

The key point here is that every observation direction (θ,ϕ) is associated with one or two plane waves heading exactly in that direction. Those plane waves headed into medium 3 (i.e. $\pi/2 < \theta < \pi$) are related to the angle β by Snell's law:

$$\sin\beta = \frac{n_3}{n_1} \sin\theta = v_3 \tag{4.10}$$

Those plane waves headed into medium 1 (i.e. $0 < \theta < \pi/2$) are either direct or related to the angle β by the law of reflection:

$$\sin\beta = \sin\theta = v_1 \tag{4.11}$$

The variable v is always the sine of the plane-wave incidence angle; it just has a different relationship with θ in medium 1 compared to that in medium 3.

The amplitude and phase of each of those plane waves is determined by sin β in combination with the three-layer Fresnel coefficients $r^{p,s}$ and $t^{p,s}$ for s- and p-pol (derivable in the same manner as equations (3.12) and (3.13); see Born and Wolf (1999)).

$$
\begin{aligned}
r^{p,s} &= \frac{r_{12}^{p,s} + r_{23}^{p,s} e^{i2k_1h(\varepsilon_2/\varepsilon_1 - v^2)^{1/2}}}{1 + r_{12}^{p,s} r_{23}^{p,s} e^{i2k_1h(\varepsilon_2/\varepsilon_1 - v^2)^{1/2}}} \\
t^{p,s} &= \frac{t_{12}^{p,s} t_{23}^{p,s} e^{i2k_1h(\varepsilon_2/\varepsilon_1 - v^2)^{1/2}}}{1 + r_{12}^{p,s} r_{23}^{p,s} e^{i2k_1h(\varepsilon_2/\varepsilon_1 - v^2)^{1/2}}},
\end{aligned}
\tag{4.12}
$$

where $t_{ij}^{p,s}$ and $r_{ij}^{p,s}$ are the single-interface Fresnel coefficients in equations (2.7–2.10). The resulting transmitted, reflected, and refracted plane waves are then reintegrated with the same weightings $\breve{\tilde{\mathbf{E}}}(\mathbf{k}_1)$ as the original plane-wave components of equation (4.3). Physically, this integration has a meaning. The amplitude and phase of each plane wave are determined by its heading, i.e. its 'ray' emanating from the dipole position in the direction (θ,ϕ), as described above. However, each plane wave is not just a 'ray.' Each is infinite in extent and is present at all observation positions, even positions other than in the path of its ray. Therefore, the total electric field $\breve{\mathbf{E}}$ at any position $\mathbf{r}$ is an integral over all the plane waves. (The details can be found in Hellen and Axelrod (1987), which, in turn, is based upon Ford and Weber (1984)).

Here are the results for regions 1 and 3, separated according to cases in which the dipole is oriented strictly in the $\hat{\mathbf{w}}$, $\hat{\mathbf{s}}$, or $\hat{\mathbf{z}}$ directions. We define $\alpha \equiv [1 - (\varepsilon_3/\varepsilon_1)\sin^2\theta]^{1/2}$. These results are only valid for observation distances $r >> z'$, which is the only practical case anyway.

For $\boldsymbol{\mu} = \mu\hat{\mathbf{w}}$:

$$
\begin{aligned}
\breve{\mathbf{E}}_1^w &= \frac{\mu k_1{}^2 e^{ik_1r}}{4\pi\varepsilon_1 r}\left(e^{-ik_1z'\cos\theta} - r^p e^{ik_1z'\cos\theta}\right)\cos\theta(\cos\theta\hat{\mathbf{w}} - \sin\theta\hat{\mathbf{z}}) \\
&= A_1(\cos\theta\hat{\mathbf{w}} - \sin\theta\hat{\mathbf{z}}) \\
\breve{\mathbf{E}}_3^w &= \frac{\mu k_3{}^2 e^{ik_3r}}{4\pi(\varepsilon_1\varepsilon_3)^{1/2} r} t^p e^{(ik_3h\cos\theta)} e^{(ik_1\alpha z')}\cos\theta(\cos\theta\hat{\mathbf{w}} - \sin\theta\hat{\mathbf{z}}) \\
&= A_3(\cos\theta\hat{\mathbf{w}} - \sin\theta\hat{\mathbf{z}}).
\end{aligned}
\tag{4.13}
$$

For $\boldsymbol{\mu} = \mu\hat{\mathbf{s}}$:

$$
\begin{aligned}
\breve{\mathbf{E}}_1^s &= \frac{\mu k_1{}^2 e^{ik_1r}}{4\pi\varepsilon_1 r}\left[e^{-ik_1z'\cos\theta} + r^s e^{ik_1z'\cos\theta}\right]\hat{\mathbf{s}} = B_1\hat{\mathbf{s}} \\
\breve{\mathbf{E}}_3^s &= -\frac{\mu k_3{}^2 e^{ik_3r}}{4\pi(\varepsilon_1\varepsilon_3)^{1/2}\alpha r} t^s e^{-ik_3h\cos\theta} e^{ik_1\alpha z'}\cos\theta\hat{\mathbf{s}} = B_3\hat{\mathbf{s}}.
\end{aligned}
\tag{4.14}
$$

For $\boldsymbol{\mu} = \mu\hat{\mathbf{z}}$:

$$\begin{aligned}\breve{\mathbf{E}}_1^z &= \frac{\mu k_1{}^2 e^{ik_1 r}}{4\pi\varepsilon_1 r}\left(e^{-ik_1 z'\cos\theta} + r^p e^{ik_1 z'\cos\theta}\right)\sin\theta(-\cos\theta\hat{\mathbf{w}} + \sin\theta\hat{\mathbf{z}})\\ &= C_1(-\cos\theta\hat{\mathbf{w}} + \sin\theta\hat{\mathbf{z}})\\ \breve{\mathbf{E}}_3^z &= \frac{\mu k_3{}^2 e^{ik_3 r}}{4\pi\varepsilon_1 \alpha r} t^p e^{(ik_3 h\cos\theta)} e^{(ik_1\alpha z')}\cos\theta\sin\theta(-\cos\theta\hat{\mathbf{w}} - \sin\theta\hat{\mathbf{z}})\\ &= C_3(-\cos\theta\hat{\mathbf{w}} + \sin\theta\hat{\mathbf{z}}).\end{aligned} \tag{4.15}$$

The scalar factors *A, B,* and *C* are abbreviations used to feature the vectorial factors that multiply them.

To verify that these fields represent, at each observation point, a transverse propagating wave heading in the direction of the ray vector **r**, we must show that the scalar product of the field vector with $\mathbf{r} = \sin\theta\hat{\mathbf{w}} + \cos\theta\hat{\mathbf{z}}$ is zero. This is clearly true for the $\boldsymbol{\mu} = \mu\hat{\mathbf{s}}$ fields (equation 4.11) because $\hat{\mathbf{s}}\cdot\mathbf{r} = 0$, and it is true for $\boldsymbol{\mu} = \mu\hat{\mathbf{w}}$ and $\boldsymbol{\mu} = \mu\hat{\mathbf{z}}$ because the resulting products cancel.

4.3 Polarization of the emitted electric field

A dipole with an arbitrary (θ',ϕ') orientation produces an electric field that is the vector sum of the fields produced by each of the dipole components:

$$\breve{\mathbf{E}}_{1,3}(\mathbf{r}, z', \theta', \phi') = \sin\theta'\cos(\phi' - \phi)\breve{\mathbf{E}}_{1,3}^{w} + \sin\theta'\sin(\phi' - \phi)\breve{\mathbf{E}}_{1,3}^{s} + \cos\theta'\breve{\mathbf{E}}_{1,3}^{z}. \tag{4.16}$$

Since each vector term on the right-hand side of equation (4.16) is orthogonal to **r**, the linear combination that forms $\breve{\mathbf{E}}_{1,3}$ is also orthogonal to **r**; i.e. transverse to the propagation direction, as it must be for propagating light. Equation (4.16) can be rewritten to separate out and combine each of the $(\hat{\mathbf{w}}, \hat{\mathbf{s}}, \hat{\mathbf{z}})$ components and expressed in vector column form:

$$\begin{pmatrix}\hat{\mathbf{w}}\cdot\breve{\mathbf{E}}_{1,3}\\ \hat{\mathbf{s}}\cdot\breve{\mathbf{E}}_{1,3}\\ \hat{\mathbf{z}}\cdot\breve{\mathbf{E}}_{1,3}\end{pmatrix} = \begin{pmatrix}\cos\theta\sin\theta'\cos(\phi' - \phi)A_{1,3} - \cos\theta\cos\theta' C_{1,3}\\ \sin\theta'\sin(\phi' - \phi)B_{1.3}\\ -\sin\theta\sin\theta'\cos(\phi' - \phi)A_{1,3} + \sin\theta\cos\theta' C_{1,3}\end{pmatrix}. \tag{4.17}$$

As seen looking back toward the dipole from the observation point, $\breve{\mathbf{E}}_{1,3}$ is composed of the sum of two orthogonal polarizations: one perpendicular to the plane of observation (s-pol) and the other residing in the plane of observation (p-pol). The s-component is already the complete s-pol; the w- and *z*- components combine to produce the p-pol. To express $\breve{\mathbf{E}}_{1,3}$ in terms of its s-pol and p-pol components, we need to rotate the coordinate axes to a new system, $(\hat{\mathbf{r}}, \hat{\mathbf{s}}, \hat{\mathbf{p}})$. The new system is just a rotation of the $(\hat{\mathbf{w}}, \hat{\mathbf{s}}, \hat{\mathbf{z}})$ system around $\hat{\mathbf{s}}$ by $(\pi/2 - \theta)$, where θ is the polar angle of the observation point. To perform this transformation, we multiply the column vector for $\breve{\mathbf{E}}_{1,3}$(equation (4.17)) by the appropriate rotation matrix:

$$\begin{pmatrix} \hat{\mathbf{r}}\cdot\check{\mathbf{E}}_{1,3} \\ \hat{\mathbf{s}}\cdot\check{\mathbf{E}}_{1,3} \\ \hat{\mathbf{p}}\cdot\check{\mathbf{E}}_{1,3} \end{pmatrix} = \begin{pmatrix} \sin\theta & 0 & \cos\theta \\ 0 & 1 & 0 \\ -\cos\theta & 0 & \sin\theta \end{pmatrix} \begin{pmatrix} \cos\theta\sin\theta'\cos(\phi'-\phi)A_{1,3} - \cos\theta\cos\theta' C_{1,3} \\ \sin\theta'\sin(\phi'-\phi)B_{1,3} \\ -\sin\theta\sin\theta'\cos(\phi'-\phi)A_{1,3} + \sin\theta\cos\theta' C_{1,3} \end{pmatrix}$$
$$= \begin{pmatrix} 0 \\ \sin\theta'\sin(\phi'-\phi)B_{1,3} \\ -\sin\theta'\cos(\phi'-\phi)A_{1,3} + \cos\theta' C_{1,3} \end{pmatrix}. \tag{4.18}$$

4.4 Emitted intensity and total power

We want to convert the electric field $\check{\mathbf{E}}_{1,3}(\mathbf{r}, z', \theta', \phi')$ into the corresponding radiated intensity $S(\mathbf{r}, z', \theta', \phi')$, because this is what is generally observed by photodetectors. S is the scalar magnitude of the Poynting vector:

$$S_{1,3} = \frac{cn_{1,3}}{8\pi} \mid \check{\mathbf{E}}_{1,3} \mid^2. \tag{4.19}$$

Even in the simplest case of a homogeneous surrounding, S clearly depends on the surrounding's refractive index (because of the prefactor). In the proximity of an interface, the situation is more complicated. Physically, the complication arises because the induced oscillating polarization in the discontinuous media affects both the radiating pattern and the dipole itself. In the special case in which medium 3 (or medium 2 if a film is present) is a metal, the radiated energy is both partially reflected and partially absorbed and converted to heat. One might initially expect that the total power, the closed integral of S around a sphere surrounding the dipole, would be a constant, i.e. independent of dipole position and orientation. However, this is not the case. The total power depends on all of the parameters $(\mathbf{r}, z', \theta', \phi')$, abbreviated here as $(\mathbf{r},\mathbf{r}')$.

Up until this point, we have assumed that the fluorophore, modeled as an emitting dipole, has a fixed amplitude of $|\boldsymbol{\mu}|$. However, in an environment where some of the emitted energy can be reflected and thereby affect the dipole, and where some of the emitted energy can be lost as heat, it does not make sense to assume that $|\boldsymbol{\mu}|$ is fixed. Instead, we need to invoke conservation of energy: the total power emitted by a fluorophore (light + heat) can be no different from the total power it absorbs from an excitation field. Continuously illuminated fluorophores should be modeled by fixed-power dipoles, not a fixed amplitude. Hence, we are interested in the radiated intensity per unit of absorbed power for the dipole:

$$\hat{S}(\mathbf{r}, \mathbf{r}') \equiv S(\mathbf{r}, \mathbf{r}')/P_T(\mathbf{r}'). \tag{4.20}$$

To calculate $P_T(\mathbf{r}')$, one possibility would be to integrate equation (4.19) over a closed sphere. However, since the field is algebraically complicated and not spherically symmetric, this approach can be arduous. Alternatively, we can start with the expression

$$P_T(\mathbf{r}') = \frac{\omega}{2}\mathrm{Im}(\boldsymbol{\mu}^* \cdot \mathbf{E}), \tag{4.21}$$

where **E** (here) is the direct field produced by the dipole at the location of the dipole, plus the reflection it produces from the surface, again at the location of the dipole. This expression is derived from the idea that the power imparted to an object moving at velocity **v** due to a force **F** is **vF**, and an oscillating dipole can be viewed as an oscillating current of moving charges. If **μ** and **E** (which both oscillate at the radial frequency ω) are relatively either in phase or exactly 180° out of phase, then $P_T(\mathbf{r}') = 0$, because in this case, the scalar product is entirely real. If **μ** and **E** oscillate at some intermediate relative phase, this is because the system radiates and/or dissipates energy.

Ford and Weber (1984) started with equation (4.21) and showed that, in the presence of a planar interface, $P_T(\mathbf{r}')$ can be written as:

$$P_T(z', \theta') = \frac{c{k_1}^4}{2{\varepsilon_1}^{3/2}}\mathrm{Re}\left\{\int_0^\infty dv \frac{v}{(1 - v^2)^{1/2}}\left[{\mu_\perp}^2 v^2(1 + \Psi^p) + \frac{{\mu_\parallel}^2}{2}(1 + \Psi^s) + \frac{{\mu_\parallel}^2}{2}(1 - v^2)(1 - \Psi^p)\right]\right\}, \tag{4.22}$$

where the abbreviation

$$\Psi^{p,\,s} \equiv r^{p,\,s}\exp\left[2ik_1 z'(1 - v^2)^{1/2}\right] \tag{4.23}$$

and the variables $r^{p,s}$ are the Fresnel reflection coefficients given in equation (4.12), which include the possibility of a thin film of medium 2 at the interface. The integration variable v can be treated as a dummy, but it is still the sine of the incidence angle. The integration extends all the way to infinity rather than 1 because $v > 1$ corresponds to evanescent emitted plane waves.

Equation (4.22) can be written in the form

$$P_T(z', \theta') = \cos^2\theta' P_T^\perp(z') + \sin^2\theta' P_T^\parallel(z'), \tag{4.24}$$

where $P_T^{\perp,\,\parallel}$ are the total powers dissipated by dipoles of fixed amplitude μ oriented perpendicular ($\theta' = 0$) or parallel ($\theta' = \pi/2$) to the interface, found by setting $\mu_\parallel = 0$ or $\mu_\perp = 0$, respectively, in equation (4.22). Figure 4.3 shows the relative $P_T^\perp$, $P_T^\parallel$ (in arbitrary units), and their ratio

$$\eta \equiv \frac{P_T^\parallel}{P_T^\perp} \tag{4.25}$$

for a glass (fused silica, $n_3 = 1.46$) : water ($n_1 = 1.33$) interface (with no film) as calculated from equation (4.22) by numerical integration using a computer.

Figure 4.3 shows that as a fixed-amplitude dipole approaches the surface, its total emitted power increases. This is why the emitted power must be normalized to the total absorbed power. The effect is orientation-dependent, and for perpendicularly oriented dipoles the effect can be an increase of about 25% or even more for a higher-refractive-index glass. This effect would not be observable as an increase in

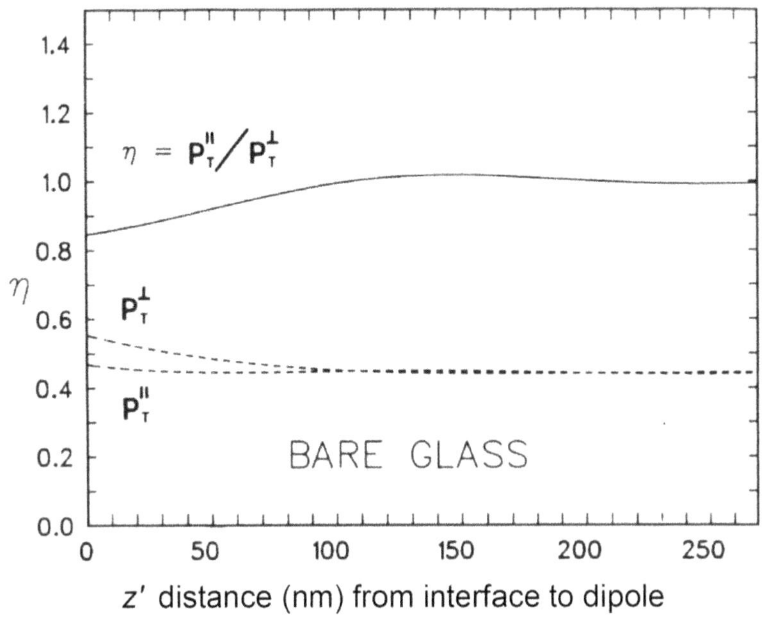

Figure 4.3. Total emitted power for the two dipole orientations relative to the substrate (parallel and perpendicular, arbitrary units) and their ratio; all are expressed as a function of the distance to the substrate. The parameters assumed are $\lambda_o = 520$ nm, $n_1 = 1.33$ (water), $n_3 = 1.46$ (fused silica), no film. Reprinted with permission from Hellen and Axelrod 1987, copyright (1987) Optica Publishing Group.

observed brightness, since the absorbed power is constant. Rather, it would be observable as a decrease in the fluorescent lifetime.

The interface draws more emitted power into medium 3 at the expense of medium 1. The observed emission on the medium 1 side, which had the advantage for dipoles located far from the interface (because of reflection back into medium 1 from the surface) actually decreases as the dipole approaches $z' = 0$. The observable emission into medium 3 thereby increases by more than 25% (for $P_T^\perp$).

4.5 Emitted intensity vs polar angle

Based on the equations given in the previous section, we can display an explicit expression for the intensity $\hat{S}(\mathbf{r}, z', \theta', \phi')$ for a single fixed-power dipole at a particular position and orientation:

$$\hat{S}_{1,3}(\mathbf{r}, z', \theta', \phi') = \frac{cn_{1,3}\left| \sin\theta' \cos(\phi' - \phi)\breve{\mathbf{E}}_{1,3}^{W} + \sin\theta' \sin(\phi' - \phi)\breve{\mathbf{E}}_{1,3}^{S} + \cos\theta' \breve{\mathbf{E}}_{1,3}^{Z} \right|^2}{8\pi\left[\cos^2\theta' P_T^\perp(z') + \sin^2\theta' P_T^\parallel(z')\right]}. \tag{4.26}$$

Figure 4.4 shows the polar observation angle θ's dependence on $\hat{S}$ (at a constant radius >> wavelength) for the case of $z' = 0$, with the pattern averaged

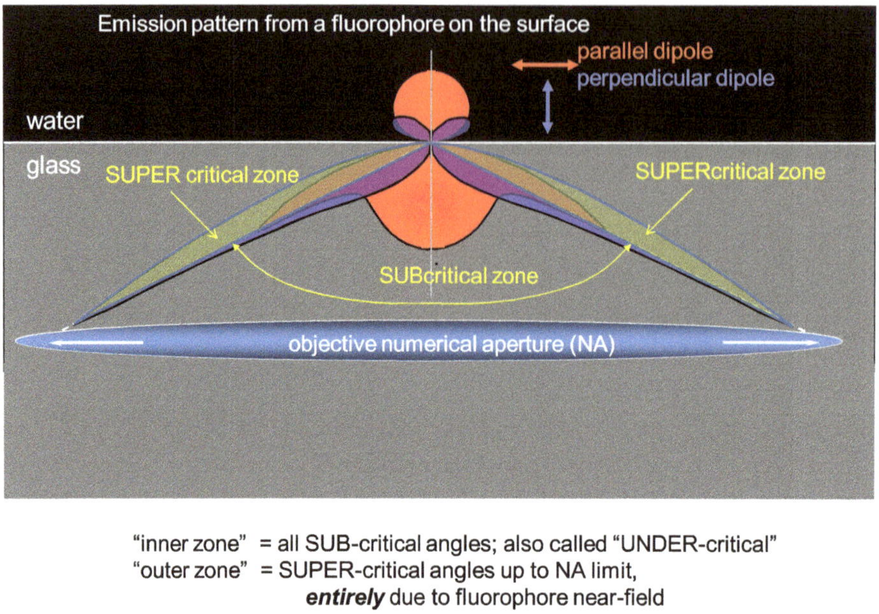

Figure 4.4. At any particular polar angle, the intensity $\hat{S}$ is proportional to the distance from the dipole (at the center) to the outer edge of the colored region in this schematic drawing; $n_1 = 1.33$, $n_3 = 1.48$. Reprinted from Axelrod (2014), copyright (2014), with permission from Elsevier.

over all azimuthal angles, for perpendicular (blue) vs parallel (red) dipole orientations. The patterns for the two orientations are quite different. The overlap region is shown in purple. Both show strong hollow-cone patterns in medium 3, especially for the perpendicular orientation, in which almost all of the intensity is in the cone wall.

This rather remarkable pattern has several important consequences for the collection of fluorescence by an objective, as listed in the following lettered paragraphs.

(a) Multiplying this fixed-power dipole intensity by the actual input power gives the actual intensity at any location in medium 1 or medium 3. If the input power decays exponentially in z' (as it would for TIR illumination in a microscope), the product is *not* exactly exponential in z'. (Exactly how it differs from exponential depends on the observation location.) However, the usual TIRF assumption that a fluorophore's observed intensity drops exponentially as it moves away from the surface is clearly not accurate.

(b) The angular pattern is a strong function of z'. Figure 4.5 shows polar angle patterns for $z' = 0$, 100, 500, 1000, and 5000 nm, all with $\lambda_o = 500$ nm and a random orientational distribution of dipoles:

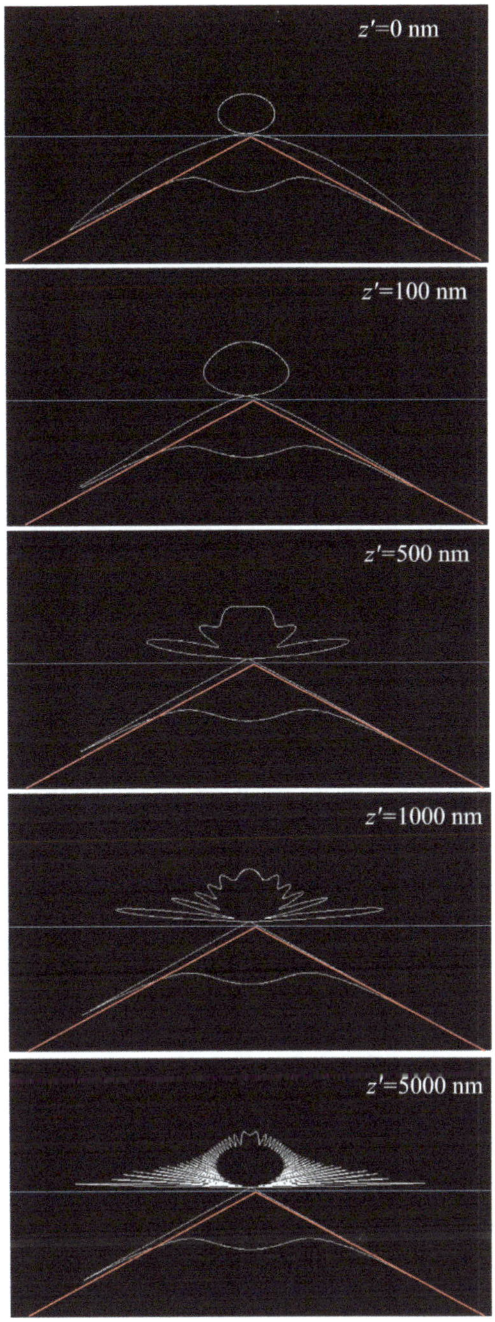

Figure 4.5. Intensity pattern of a dipole at a surface (blue line). The intensity at any particular polar angle is proportional to the graphed distance from the dipole to the white line. The red line is at the maximum polar angle θ_c at which light propagating from medium 1 can refract into medium 3. Any light that appears at a higher polar angle in medium 3 originates from evanescent (near-field) emission from the dipole. In medium 3, this zone is called 'supercritical.' Parameters assumed: n_1 = 1.33, n_3 = 1.515, λ_o = 500 nm.

Supercritical-emission light only appears to a substantial extent in medium 3 for dipoles that are very close (i.e. subwavelength z') to the surface. At any particular supercritical polar angle, the intensity varies as an exponential decay with z'; the larger the angle, the more rapid the characteristic decay. However, if a range of supercritical angles are observed together, the decay with z' is not a single exponential but rather a weighted sum of exponentials.

In medium 1, the rapid variations with angle (with increasing rapidity as z' increases) are due to interference between direct and reflected light. However, these diagrams assume that the emission is monochromatic (i.e. coherent). Fluorescence generally has a significant wavelength spread and the interference fringes become blurred out for z' greater than the coherence length (a few wavelengths).

For small z', most of the emission goes into medium 3 because the dipole's near field interacts with the surface and gets converted to propagating light in medium 3. But for large z', the dipole's near field is out of range of the surface, and medium 1 then inherits most of the emission because some of it reflects at the surface. However, as can be seen in figure 4.5, much of that medium 1 light propagates at too high a polar angle to be captured by an objective. In general, observing the fluorescence through medium 3 (i.e. a coverslip) is more efficient, especially for fluorophores near the surface.

(c) An objective with a high enough numerical aperture to subtend a capture angle larger than θ_c can capture the supercritical emission. However, if the fluorophores being observed are not near the surface, then a high-aperture objective has no advantage in light gathering over one that captures just up to θ_c.

(d) Even without TIR excitation (which is designed to preferentially excite surface-proximal fluorophores), a high-aperture objective exhibits a z' sensitivity that somewhat favors surface-proximal fluorophores because of near-field capture ability.

(e) The ability of an objective to capture light in medium 3 is clearly *not* a linear function of its subtended solid angle, and the obtainable resolution is more complicated than the standard inverse 'Raleigh resolution limit' relationship with numerical aperture, because of the complicated intensity vs polar angle profile, as well as the possible phase variations of the emitted electric field in different directions. The topic of resolution, including these near-field effects, will be covered in section 4.10.

4.6 Total fluorescence collection through a microscope objective

4.6.1 Single dipole: integration over azimuthal angles

A microscope objective implicitly integrates the intensity pattern over all 2π azimuthal angles and over all polar angles subsumed by the objective's numerical aperture. The ϕ-averaged intensity obtained from a fixed-power dipole is:

$$\langle \hat{S}_{1,3}(z', \theta, \theta') \rangle_{\phi} = \frac{1}{2\pi} \int_0^{2\pi} d\phi \langle \hat{S}_{1,3} \rangle. \tag{4.27}$$

Performing this operation on equation (4.26), and noting that all the cross terms implicit in the square integrate to zero, we get:

$$\langle \hat{S}_{1,3}(z', \theta, \theta')\rangle_{\phi} = \frac{\langle \hat{S}_{1,3}\rangle_{\phi}^{\perp}(z',\theta)}{1 + \eta(z')\tan^2\theta'} + \frac{\langle \hat{S}_{1,3}\rangle_{\phi}^{\parallel}(z', \theta)}{1 + [\eta(z')\tan^2\theta']^{-1}}, \tag{4.28}$$

where

$$\begin{aligned} \langle \hat{S}_{1,3}\rangle_{\phi}^{\perp}(z', \theta) &\equiv \frac{cn_{1,3}}{8\pi}\frac{\left|\breve{\mathbf{E}}_{1,3}^{s}\right|^2}{P_T^{\perp}(z')} \\ \langle \hat{S}_{1,3}\rangle_{\phi}^{\parallel}(z', \theta) &\equiv \frac{cn_{1,3}}{16\pi}\left[\frac{\left|\breve{\mathbf{E}}_{1,3}^{w}\right|^2 + \left|\breve{\mathbf{E}}_{1,3}^{z}\right|^2}{P_T^{\parallel}(z')}\right]. \end{aligned} \tag{4.29}$$

4.6.2 Single dipole: integration over polar angles

The fraction Q of the total energy dissipated by a fixed-power dipole and collected by a microscope objective centered around the normal to the interface is obtained by integrating the *non*-normalized intensity S over the observation angles subsumed by the objective, and then dividing this by P_T:

$$\begin{aligned} Q_{1,3}(z', \theta') &= \frac{r^2\int S_{1,3}\sin\theta\, d\theta d\phi}{P_T(z', \theta')} = r^2\int 2\pi\langle \hat{S}_{1,3}\rangle_{\phi}(z', \theta)\sin\theta\, d\theta \\ &= \left[\frac{Q_{1,3}^{\perp}(z')}{1 + \eta(z')\tan^2\theta'} + \frac{Q_{1.3}^{\parallel}(z')}{1 + [\eta(z')\tan^2\theta']^{-1}}\right], \end{aligned} \tag{4.30}$$

where

$$Q_{1,3}^{\perp,\parallel}(z') \equiv 2\pi r^2\int \langle \hat{S}_{1.3}\rangle_{\phi}^{\perp,\parallel}(z', \theta)\sin\theta\, d\theta. \tag{4.31}$$

The limits of integration are related to the maximum polar angle $\alpha_{1,3}$ subtended by the objective's numerical aperture NA, where NA $= n_{1,3}\sin(\alpha_{1,3})$. Note that the observation polar angle θ is always measured from the positive direction of the z-axis, so $\alpha_1 = \theta$ for an objective looking down in medium 1 and $\alpha_3 = \pi{-}\theta$ for an objective looking up in medium 3. Therefore, the limits of θ integration in medium 1 are $(0,\alpha_1)$, and in medium 3 are $(\pi, \pi{-}\alpha_3)$.

The parameter Q is crucial to understanding the efficiency with which an objective gathers light from a fluorophore. Figure 4.6 (left panel) shows Q as a function of fluorophore orientation, distance from the surface, and direction of viewing. It clearly shows that, for an objective with a high enough NA to gather

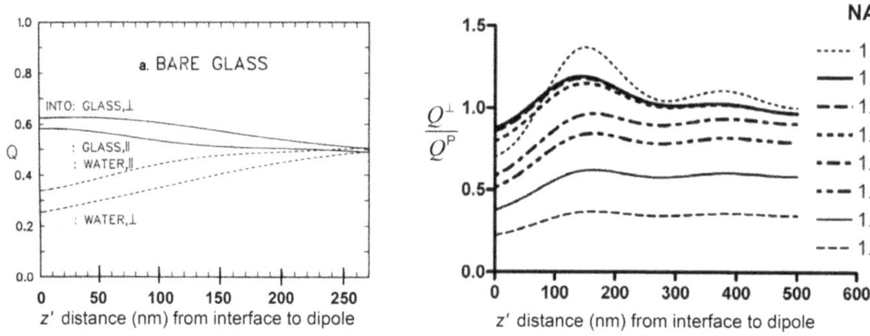

Figure 4.6 (Left panel) Collection efficiency Q vs. z' distance for NA = 1.4, λ_o = 520 nm, and n_3(fused silica glass) = 1.46. Reprinted with permission from Hellen and Axelrod 1987, copyright (1987) Optica Publishing Group. (Right panel) The ratio $Q^{\perp}/Q^{\parallel}$ (viewing through the glass coverslip) vs. z' distance. The coverslip is assumed to have n_3 = 1.52 except for the NA = 1.65 objective, which requires a special coverslip with n_3 = 1.78. Note that the 1.49 objective shows the least deviation from unity over the whole range of z' distances.

some of the subcritical near-field light, viewing from below through the coverslip is about twice as efficient as viewing from above through the water.

4.6.3 Distribution of dipoles

To calculate the fluorescence power observed through a microscope objective from a collection of fluorophores at one location, we must perform three operations on equation (4.26): (a) integration over all azimuthal angles of observation ϕ (0 to 2π) ; (b) integration over those polar angles of observation θ that can be captured by the objective's numerical aperture; and (c) specification of the excitation polarization and orientational distribution of fluorophores.

For an arbitrary distribution C (in z' position and orientation) of dipole concentration along the optical axis (i.e., x'=y=0), the observed total observed fluorescence F is:

$$F = \xi \int C(z', \theta', \phi') \mid \boldsymbol{\mu}^{ab} \cdot \mathbf{E}^{ex} \mid^2 Q(z', \theta') \sin\theta' d\theta' d\phi' dz', \tag{4.32}$$

where the scalar product squared term is proportional to the probability of excitation of an absorption dipole $\boldsymbol{\mu}^{ab}(\theta',\phi')$ by a (possibly evanescent, possibly polarized) incident electric field $\mathbf{E}^{ex}(\theta',\phi')$, and the constant ξ is a product of all relevant conversion factors, normalizations, and arithmetic constants. Of course, the solid angle subtended by the objective slowly decreases with increasing z'; we consider here only a range of z' over which the subtended solid angle is essentially constant. The depth of a TIR evanescent field (~0.1 μm) and even the depth of focus in EPI illumination (several μm) are well within this range.

In a common special case, distribution C is assumed to have random orientations throughout its z' range:

$$C(z', \theta', \phi') = C(z') \tag{4.33}$$

and the incident electric field is assumed to exponentially decay in z', as in TIR illumination. A technique called 'variable-angle TIRF' has been developed with the goal of determining the form of $C(z')$. The exponential decay depth d of the TIRF field is varied by systematically stepping through a range of incidence angles (see equation 2.12), and the following simple formula is assumed:

$$F(d) = \xi E_0{}^2 \int C(z')e^{-z'/d}dz'. \tag{4.34}$$

Using an experimental estimate of $F(d)$, the concentration profile $C(z')$ can (in principle) be computed as the inverse Laplace transform of $F(d)$. For a finite number of data points $F(d)$, the computation can present uniqueness problems. However, in light of equation (4.31), the whole procedure must be viewed as an approximation, even if it were computable, because of the complicating presence of Q in the integrand. The problem is that different z' positions and different orientations in C have different P_T normalizations, so equation (4.32) is a more accurate starting point than equation (4.34).

How big an effect does near-field interaction with the surface—and consequent near-field collection by an objective—actually have on the 'exponential decay' of intensity as a fluorophore is moved away from the surface? Figure 4.7 shows a particular example from computer calculations based on equation (4.32) with $C(z', \theta', \phi') = C_0\delta(z' - Z)$, in which the dipole distribution is located entirely at a single $z' = Z$, but randomly oriented. The black curve represents the collection efficiency, Q, and the numbers on the

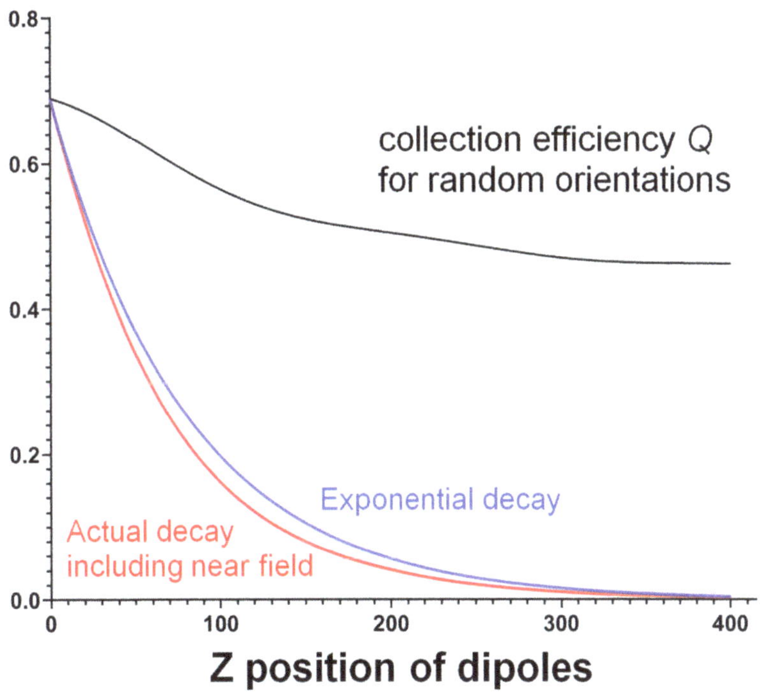

Figure 4.7. Emission intensity collected from a randomly oriented distribution of excited fluorophores by an objective vs the distance z' between the fluorophores and the surface. The parameters assumed here are NA = 1.49, evanescent field depth d = 80 nm, n_1 = 1.33 (water), n_3 = 1.515 (BK-7 glass), no intermediate film, λ_o = 500 nm.

vertical axis correspond to its (unitless) numerical value. The blue curve represents TIR-excited fluorescence as a pure exponential decay, i.e. ignoring all the special optical effects of surface proximity upon emission. The red curve represents the actual observed TIR fluorescence, including all the effects discussed in these sections, summarized by equation (4.32). The two fluorescence curves are normalized to the same value at $Z = 0$ to emphasize the slightly shorter decay distance of the actual TIR fluorescence. In fact, near-field capture makes the actual collected fluorescence brighter than it otherwise would be (i.e. what it would be for dipoles far from the surface) by a factor of about 1.54 for the parameters chosen. The 'excess' numerical aperture (>1.33) serves only to gather supercritical near-field light. Even with z-independent EPI illumination, there is a bias toward the collection of light from near-surface fluorophores, as shown by the Q vs z curve.

4.7 Pattern at the back focal plane

As discussed above, some of the near-field light emitted from a fluorophore is converted into a hollow cone of propagating light in a nearby glass substrate. This hollow cone can then be captured by a high-aperture objective, as long as its NA is greater than the refractive index of the medium in which the fluorophore resides

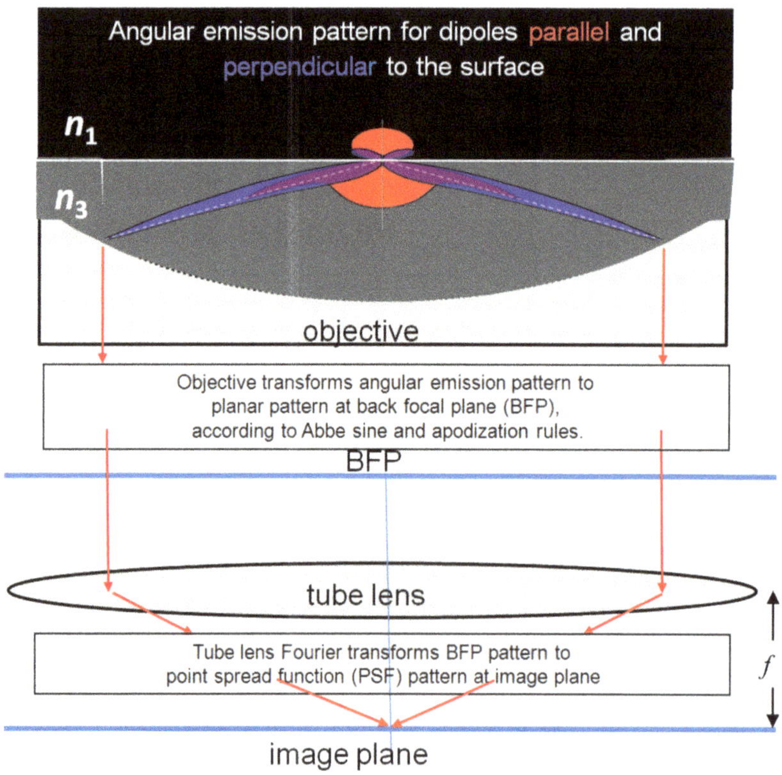

Figure 4.8. Schematic view of the path of near-field light in a microscope from supercritical angles in medium 3, then through the objective, BFP, and tube lens, and toward a focus as a point spread function (PSF) in the image plane.

(e.g. 1.33 for water). Since the objective performs a one-to-one conversion of polar angles at which light enters the objective into radial positions at the objective's back focal plane (BFP), the hollow cone of near-field light is converted into an annulus of illumination through the BFP, with radii corresponding to NA > 1.33 (see figure 4.8). None of the far-field-propagating energy, for which $(k_{1x}^2 + k_{1y}^2) \leq k_1^2$, is cast into that high-angle hollow cone or BFP annulus; rather, it is all cast into 'subcritical' angles and a central BFP disk corresponding to NA < 1.33.

The light-collecting advantage of very high-aperture objectives (NA > 1.33) resides purely in their ability to capture supercritical near-field light. The extra numerical aperture does not help in gathering far-field emission light, because none of that light propagates in the glass at the supercritical angles. For small z, most of the light gathered by a high-aperture objective comes through the supercritical zone. The 'extra' aperture, say, in going from NA = 1.33 (for which there is no supercritical zone) to 1.49 more than doubles the objective's light-gathering ability, which is a strong reason for using the highest apertures available.

Each supercritical polar angle of emission (and radius in the BFP) corresponds to a particular near-field x–y wavefront spacing in the fluorophore near field and hence a particular z-direction exponential decay length. However, since the BFP annulus has a finite thickness containing a range of supercritical angles up to the limiting aperture of the objective, the corresponding decay lengths in this annulus also cover a range of decay distances. Therefore, the total light power coming through the BFP's supercritical annulus is not a single exponential. The association of a particular BFP radius with a particular decay length is completely analogous to the TIRF effect for light going the other way (excitation), whereby a particular angle of incidence gives rise to a particular TIR evanescent decay depth. Likewise, on the emission side, concentration profiles should also be deducible from the supercritical intensity versus BFP radius curve.

In section 4.11, we will discuss how the segregation of the near-field vs far-field light at the BFP can be utilized in experiments. As a preliminary to that, we would like to generate theoretical images of how the BFP would 'look' if it were observed, say, simply by removing the microscope eyepiece and looking down through the tube. The mathematical form of the mapping of polar angle θ into radii r in the BFP is an important matter in objective lens design. The objective ideally performs the mapping 'one-to one' while converting from a spherical to a flat surface. Therefore, the field $\breve{\mathbf{E}}_{1,3}(\mathbf{r}, z', \theta', \phi')$ calculated in equation (4.17) in spherical coordinates is also considered to be the field on the BFP plane, except for an additional 'apodization' factor $[= (1/\cos\theta)^{-1/2}]$ that accounts for foreshortening. The Abbe sine condition specifies that r must be proportional to $\sin\theta$ to preserve maximum image sharpness. On the other hand, the Abbe sine condition distorts distances in the off-axis zones of the image plane, and for that reason, r proportional to $\tan\theta$ may be better. To avoid this geometrical optics dilemma, here, we generate views of the intensity at the BFP based on equation (4.32) and display them such that r is directly proportional to θ. From a physical optics point of view, the one-to-one association between r and θ is an approximation, because it ignores the (slight) diffraction spreading of the rays between the objective and the BFP. These considerations modify the spherical field emitted by the dipole $\breve{\mathbf{E}}_{1,3}$ into the flat field $\mathbf{E}'_{\mathrm{BFP}}$ at the BFP.

The BFP pattern shown in figure 4.9 can be used to provide a sensitive quantitative reading of both the polar and azimuthal angles of the fluorophore orientation. The figure shows a computer-calculated view of the intensity pattern (with no polarizers in the path) at the BFP for the no-film case ($h = 0$) at three different fluorophore orientations ($\theta = 0, \pi/4, \pi/2$, all at $\varphi = 0$). Note that the patterns are quite distinct for different polar orientation angles, and that the azimuthal angles are unambiguous, i.e. the orientation (θ, φ) should be quite distinguishable from $(\theta, \pi + \varphi)$ if θ is intermediate between zero and $\pi/2$.

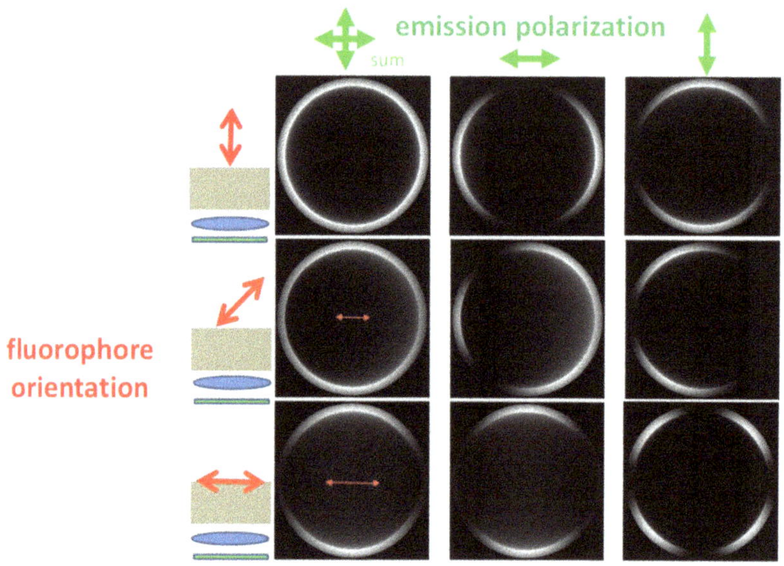

Figure 4.9 BFP intensity pattern vs single dipole/fluorophore orientation (shown in red) and emission polarization (shown in green) located on a bare TIR surface ($z = 0$) for NA = 1.49, $n_1 = 1.33$, $n_3 = 1.52$.

As expected, the loss of the ability to collect near-field emission as z' increases can be directly seen by examining the BFP, which is computer-generated for figure 4.10 based on equation (4.32), again for $z = 0$ and NA = 1.49.

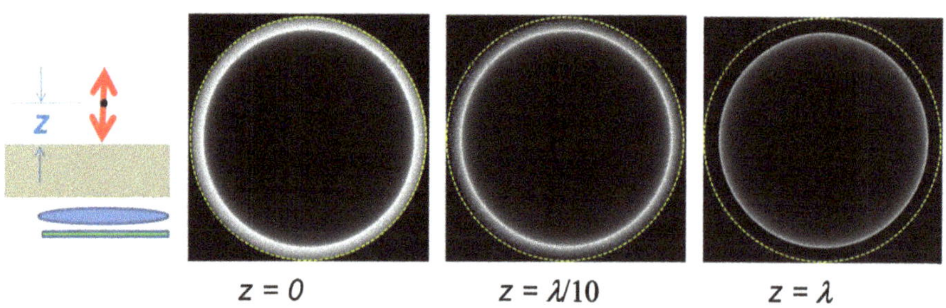

Figure 4.10. Computer-calculated view of the intensity pattern at the BFP shown in grey scale for the no-film case ($h = 0$) at three different fluorophore distances from the surface. The parameters assumed are NA = 1.49, $n_1 = 1.33$; $n_3 = 1.52$.

4.8 Characterization of films with supercritical-emission light

Apart from the unambiguous information about dipole orientation, the supercritical zone in the BFP also reports on the thickness h of an intermediate-layer film. The intensity of supercritical light captured from a fluorophore by a bare substrate and sent through supercritical radii in the BFP decreases monotonically with radius. However, the presence of a film that has an intermediate refractive index on the substrate can lead to multiple fluorescence peaks in the supercritical BFP zone. Figure 4.11 shows a theoretical reconstruction of this effect. These supercritical-emission peaks in the BFP are analogous to the TIR excitation evanescent field intensity peaks that can occur due to partial resonance in the film, as discussed earlier. For each film thickness h, the shape of the BFP intensity vs. radius curve is qualitatively unique. Therefore, it should be possible to identify h (and also the refractive index of the film) by matching an experimentally observed curve shape to a theoretically predicted curve.

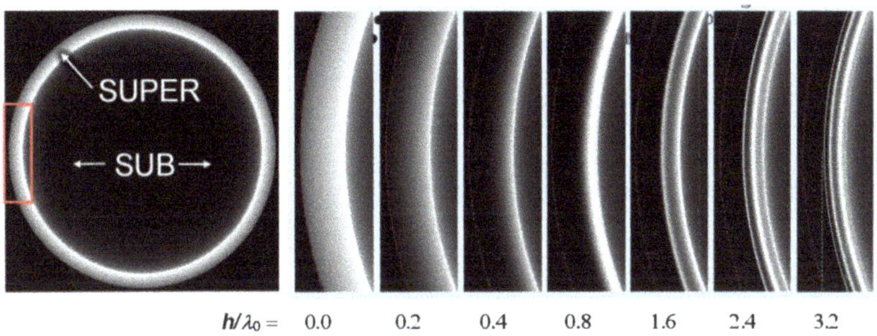

Figure 4.11. Detail of the BFP pattern in the supercritical zone for various film thicknesses, all with $(\theta',\phi') = (0,0)$ and with $z' = 0$ (the dipole resides in n_1 at the $n_1{:}n_2$ interface). The parameters assumed in these calculations are NA = 1.49, $n_1 = 1.33$; $n_2 = 1.42$; $n_3 = 1.52$. Reprinted from Axelrod 2012, copyright (2012) John Wiley & Sons.

4.9 Effect of metal films on fluorescence emission

We have already examined (section 4.3) how an intermediate film made of metal can have a strong effect on the TIR-induced evanescent field, including a great enhancement of intensity at a particular narrow range of incidence angles and suppression of evanescent intensity at all other angles (see figure 2.2). This is due to 'surface plasmon resonance' driven by the incident field. A metal film also has strong effect on emission patterns, because emission light itself can drive surface plasmon resonance in the film. This effect has some potentially important consequences for experimental applications, so we will devote this whole section to discussing how a metal film affects emission.

Qualitatively (for both excitation and emission-induced surface plasmons), the driving electric field (i.e. light) induces electrical polarization in the film, the sign of which flips sinusoidally in the lateral direction at the same spatial frequency as the driving field in the lateral direction. This spatial frequency depends upon the incidence angle of the driving field. The separation of charges consequently results in a restoring force: the charges 'want' to return to equilibrium at a uniform distribution. The restoring force is a function of the lateral spatial frequency, so it too is a function of the incidence angle. Such a system has a natural resonant frequency, because the mobile electrons have an effective mass which causes them to 'overshoot' the equilibrium. If the driving temporal frequency of the applied electric field equals that of the natural resonant frequency, then the induced polarization can become quite large, generating a significant re-emission of light.

Quantitatively, the equations for the emitted electric field (culminating in equation 4.29) predict this effect if, in equation (4.12), the Fresnel coefficients are computed for a film that has the complex refractive index typical of a metal at optical frequencies, i.e. a small positive real part and a large positive imaginary part. Apart from this special refractive-index feature of metals, the same theory applies as for dielectric films.

A metal film strongly affects the total power P_T emitted by a constant-amplitude single oscillating dipole (figure 4.12a,b).

Clearly, P_T approaches very large values near a metal film for z' approaching zero. Of course, a real-world dipole does not emit infinite energy in any finite time. But the normalization to P_T (equation 4.20) assures us that the emitted light intensity approaches zero for small z'. The output energy (which always equals the input energy) then appears in the form of heat. Metal films at proximal distances 'quench' fluorophore emission.

The quenching effect can be seen more directly, for example, by plotting (figure 4.12c) the P_T-normalized intensity $\hat{S}$ (equation 4.26) emitted at the specific polar angle $\theta = 70°$ for the case of a dipole oriented perpendicularly to the surface ($\theta' = 0$, so that azimuthal angle does not matter). $\hat{S}$ increases very rapidly from $z' = 0$ to $z' = 15$ nm (which is much shorter than the light wavelength) to about half of its ultimate maximal value at around 120 nm. Figure 4.13 expands the z' abscissa of figure 4.12c to emphasize the quenching at short z', here plotted not at a particular polar angle, but rather as the collection efficiency Q as seen through an NA = 1.4 objective located either in the water (n_1) or in the glass (n_3).

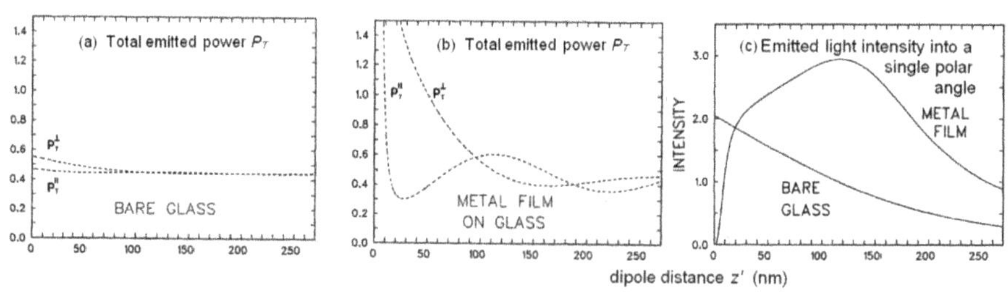

Figure 4.12. (a) Total power P_T dissipated by fixed-amplitude dipoles oriented parallel and perpendicular to the interface (dashed lines and arbitrary units) versus dipole distance z' measured into the water. The units for P_T are arbitrary. (b) The same, except for an aluminum film (h = 22 nm) on glass, rather than bare glass. (c) Light intensity emitted into an observation polar angle $\theta = 70°$ versus dipole distance z' measured into the water. For all three panels, $\lambda_0 = 520$ nm, and the refractive indices are as follows: $n_1 = 1.33$ (water); $n_2 = 0.859 + i6.08$ (aluminum); $n_3 = 1.46$ (fused silica glass).

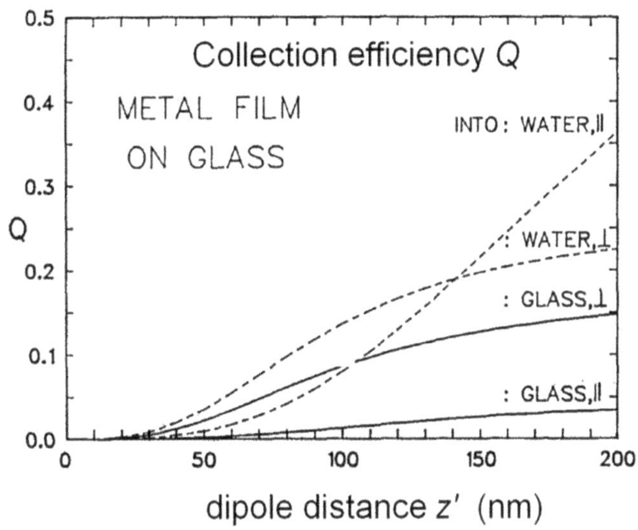

Figure 4.13. Collection efficiency Q vs dipole distance z' as viewed through an NA = 1.4 objective located either in the water (n_1) or in the glass (n_3), for a single dipole oriented either parallel or perpendicular to the surface. The parameters are all the same as for figure 4.12. Reprinted with permission from Hellen and Axelrod 1987, copyright (1987) Optica Publishing Group.

The sharp increase in Q as z' increases from zero up through its very low values has potential practical uses in axial superresolution. A phospholipid membrane bilayer supported on an aluminum-coated glass substrate, for example, might contain a fluorescent label in both leaflets. However, the fluorescence from the leaflet proximal to the substrate is almost completely quenched, so the effective view is from only one half of a phospholipid bilayer. The very high P_T near $z' = 0$ also implies that the fluorescence lifetime of fluorophores there is very short. This gives them little time to photobleach, which is an event that takes place from the excited state. In this case, a deliberate photobleaching pulse of bright light will bleach only the distal leaflet, and a recovery of fluorescence will report the rate of intra-membrane interleaflet 'flip-flop' of fluorophore. Another use of metal-film quenching is to suppress a fluorescent signal from non-specific adsorption of a labeled molecule directly to the substrate surface, while still preserving its signal from (presumably more interesting) structures in overlying cells (see section 6.3).

All of the foregoing discussion depends on the simple utilization of a complex n_2 appropriate for the metal in the equations for the Fresnel coefficients. This procedure is itself an approximation; it is no longer accurate or valid for small z' distances that are comparable with the mean free path of electrons in the metal. For aluminum, the mean free path is 19 nm. For this reason, a more accurate theory (the 'quasi-static semiclassical infinite barrier' (SCIB) was used to generate figures 4.12 and 4.13. For calculations of P_T, the SCIB model gives larger results by about 40% for $z' = 1$ nm, and by about 4% for $z' = 5$ nm.

The presence of a metal film qualitatively changes the shape of the emission pattern. Of course, this change is only relevant for very small z' distances, for which quenching dominates, zeroing out the emission in every direction. Using the same theory as that used in generating figure 4.5 for bare glass, except with an aluminum film of $h = 22$ nm, it is possible to generate the emitted intensity pattern S averaged over the azimuthal angle (figure 4.14).

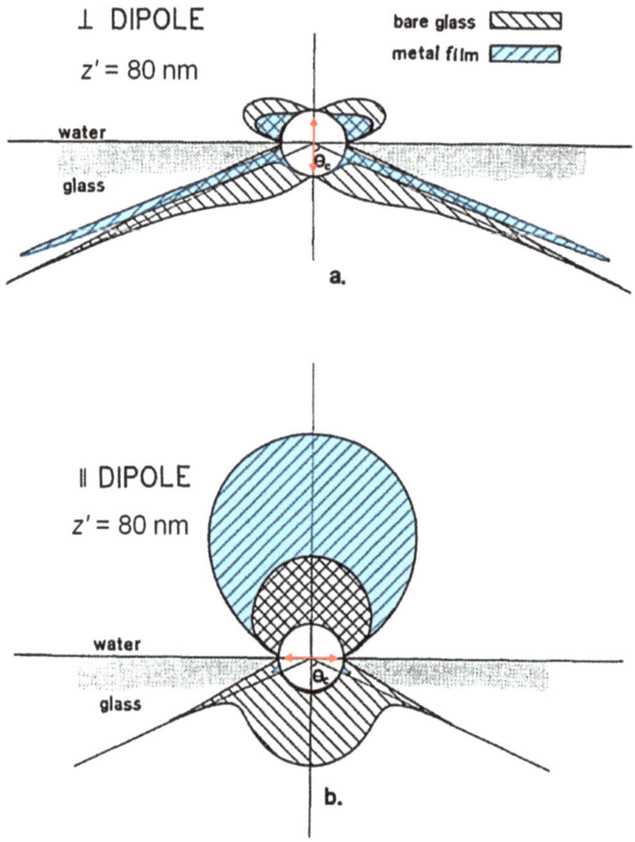

Figure 4.14. ϕ - averaged normalized intensities $\langle \hat{S} \rangle_{\phi}^{\perp,\parallel}$ versus polar observation angle for a dipole at $z' = 80$ nm from the interface. The blue/black crosshatch region denotes the pattern for an aluminum film with a thickness of $h = 22$ nm. The white/black crosshatch region denotes the pattern for a bare surface (no film). The intensities observed at each polar angle are drawn as proportional to the radial distance from the central circle. The relative dielectric constants for the three media are $\lambda_0 = 520$ nm, $\varepsilon_1 = 1.77$ (water), $\varepsilon_2 = -32.5 + 8.4i$ (aluminum), and $\varepsilon_3 = 2.31$ (glass). The corresponding refractive indices are $n_i = \varepsilon_1{}^2$. Reprinted with permission from Hellen and Axelrod 1987, copyright (1987) Optica Publishing Group.

Several interesting features in the metal-film emission pattern can be noted:

- For the perpendicular dipole orientation, a very thin-walled hollow cone of emission light is created, with a polar angle given by the 'surface plasmon angle,' given by equation (3.22). This equation was written for the surface plasmon effect in the reverse direction, i.e. setting up an intense evanescent field using incoming collimated laser illumination. But (except for the dependence of the dielectric constants on wavelength, which is generally longer for emission than excitation), the theory is the same. Most of the

emitted light actually goes into that hollow cone; i.e. through the metal film, rather than directly into the water, where there is no such obstacle.
- For a parallel dipole orientation, the hollow cone of light emitted through the metal is extremely dim. Almost all the emitted light goes into the water.
- These effects correspond to the incident light surface plasmon effects discussed in chapter 3. There, p-polarized incident light exhibited a large surface plasmon enhancement, and s-polarized light barely set up any evanescent field at all in the water.
- The surface plasmon angle is larger than the sharp emission peak seen on bare glass, a peak which occurs at the TIR critical angle for the emission wavelength.
- Classical reflection from the metal surface shows up as the much greater emission of parallel dipoles (vs. perpendicular) into the water, at least for z' much smaller than a wavelength. The film sets up an image dipole just below the film surface. In the case of parallel dipoles, the real and image dipoles oscillate in phase, and if z' is small, their emission patterns constructively interfere. However, in the case of perpendicular dipoles, the real and image dipoles oscillate out of phase and their emission patterns tend to cancel.

4.10 Pattern at the image plane

The tube lens in a microscope converts the electric field pattern at the BFP into an electric field pattern at the image plane. If the source of the light was a single emitting dipole, then the image intensity (the square of the electric field amplitude) is the PSF, i.e. how the microscope transforms an infinitesimal point into a finite-sized blur.

Just as the (negligible) diffraction in the space between the objective and the BFP was ignored, we assume that the E field present at the BFP is preserved, diffraction-free, as it impinges upon the tube lens, which focuses the light to the image plane (IP) in an infinity-corrected system. In general, even for sources that are more extended than a single dipole, the field $\mathbf{E}'_{\text{IP}}$ at the image plane of the tube lens (which is conjugate to the sample plane of the microscope) is then proportional to the Fourier transform of the field $\mathbf{E}'_{\text{BFP}}$ at the BFP:

$$\mathbf{E}'_{\text{IP}}(x_{\text{IP}}, y_{\text{IP}}) = AB \int_{-\infty}^{\infty} \mathbf{E}'_{\text{BFP}}(x, y) \exp\left[-i\frac{k}{f}(x_{\text{IP}}x + y_{\text{IP}}y)\right] dxdy \tag{4.35}$$

where (x,y) are the coordinates at the BFP; $(x_{\text{IP}}, y_{\text{IP}})$ are the coordinates at the image plane, f is the focal length of the tube lens; and $k = 2\pi/\lambda$. Factor A is a constant independent of $(x_{\text{IP}}, y_{\text{IP}})$, and factor $B = \exp[i(k/2f)(x_{\text{IP}}^2 + y_{\text{IP}}^2)]$ describes the phase curvature; neither factor has any impact on the form of the intensity $I'_{\text{IP}} = |\mathbf{E}'_{\text{IP}}|^2$ in the microscope's image plane.

4.10.1 Approximation of the PSF: the 2D Airy disk

For the (vastly over)simplified case in which fluorescent light from an infinitesimal object under microscopic view uniformly illuminates the objective across its entire aperture, the PSF is an 'Airy disk.' As shown in the sections above, the emission pattern from a point oscillating dipole is far from uniform. However, a derivation of the Airy disk's mathematical pattern and its corresponding contrast transfer function (CTF) is useful in order to understand lateral resolution in TIRF and improve it, a topic covered in chapter 6. The derivation here is based upon an 'infinity-space' microscope configuration, shown in figure 4.15.

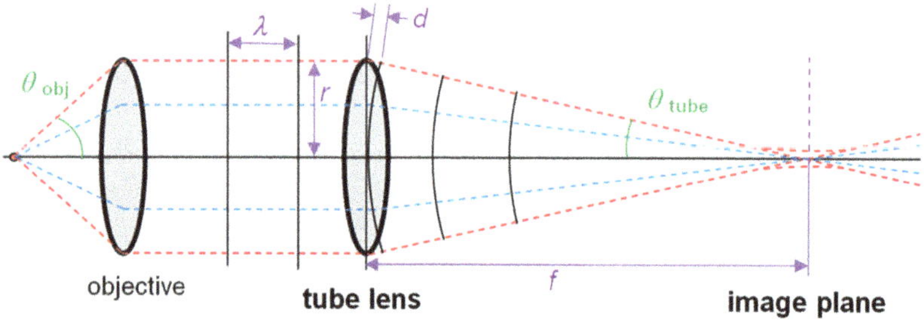

Figure 4.15. Ray and wavefront diagram of extreme rays (red) and intermediate rays (blue) from a point source to the image plane. The wavelength λ is greatly exaggerated for pictorial clarity. The radial position of a general ray is r. The relative advance of the wavefront for which $r > 0$ is d.

Consider first the tube lens, the whole function of which is to convert parallel collimated wavefronts into convergent spherical wavefronts. It performs this conversion by advancing the off-axis front at radius r by a distance d (or more precisely, by retarding the on-axis portion). The Pythagorean theorem gives

$$\begin{aligned}(d+f)^2 &= r^2 + f^2 \\ d(d+2f) &= r^2.\end{aligned} \tag{4.36}$$

In a small-angle approximation ($\theta_i << 1$), $d << 2f$, so $d = r^2/2f$. This corresponds to a phase-angle advance of $\phi = 2\pi d/\lambda = kr^2/2f$. Therefore, the curved wavefront's electric field just as it emerges from the tube lens has the form

$$E(\mathbf{r}) = E_0(\mathbf{r})\exp\left(-\frac{ikr^2}{2f}\right), \tag{4.37}$$

where $\mathbf{r}$ is a 2D radial vector in the plane of the (idealized, thin) tube lens. $E_0(\mathbf{r})$ allows for the possibility that the electric field amplitude varies on the curved wavefront in that plane.

The question is: how does this curved wavefront propagate in free space toward the focal region and what sort of pattern does it produce there? Each point on that wavefront is assumed to generate a spherical wavelet which travels outward at the speed of light ('Huygens wavelets'). In fact, these wavelets produce the wavefronts not only ahead and behind where we started: they even produce the wavefront that 'generated' them. Figure 4.16 shows an example with straight wavefronts:

The wavelets are represented by the spherical wave form e^{iks}, where k is the wavenumber ($=2\pi/\lambda$) and s is the distance from the origin of that wavelet to the point P where the wave is

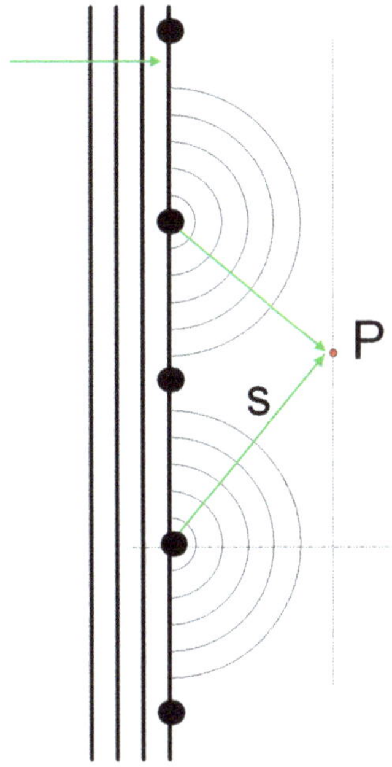

Figure 4.16. Schematic for wavelet propagation.

being reconstructed. It is reasonable that the amplitude should also be proportional to $1/s$, because then, the intensity E^*E drops according to $1/s^2$ for each wavelet.

However, the wavefronts have three other features which are not so intuitive. One is that they are phase shifted from the phase of the point 'generating' them by 90° (denoted by the multiplicative factor $1/i$ below in equation). This is necessary so that the final result is in phase with the generating wavefront and in fact that the generating wavefront can be generated correctly by its own wavelets. Second, there is a scaling factor of $1/\lambda$ which also falls out of the math for self-generation. And third, these wavelets have to account for the forward-traveling directionality of the light. It can be shown that there must be a multiplicative factor of $(1 + \cos\psi)$, where ψ is the angle between the normal to the generating wavefront and the direction toward the point P. Therefore, each wavelet has the form:

$$\frac{e^{iks}}{i\lambda s}(1 + \cos\psi). \tag{4.38}$$

In all the calculations to follow, we will simplify things by assuming that we only have small angles of convergence, so that we can ignore the ψ dependence.

First, consider the propagation of a wavelet in the free space after the tube lens (see figure 4.17). The source is some point P on the plane immediately after the tube lens. The circularly symmetric electric field on that plane is $E(\mathbf{r})$, whose position coordinates are denoted by unprimed variables. For simplicity, we ignore the fact that the electric field is a vector. We define some arbitrary plane (later to be an image plane of the tube lens) a distance L downbeam and the point P' to be at position $\mathbf{r}'$ in the 2D space on that plane. Because of the circular symmetry of the point dipole, the

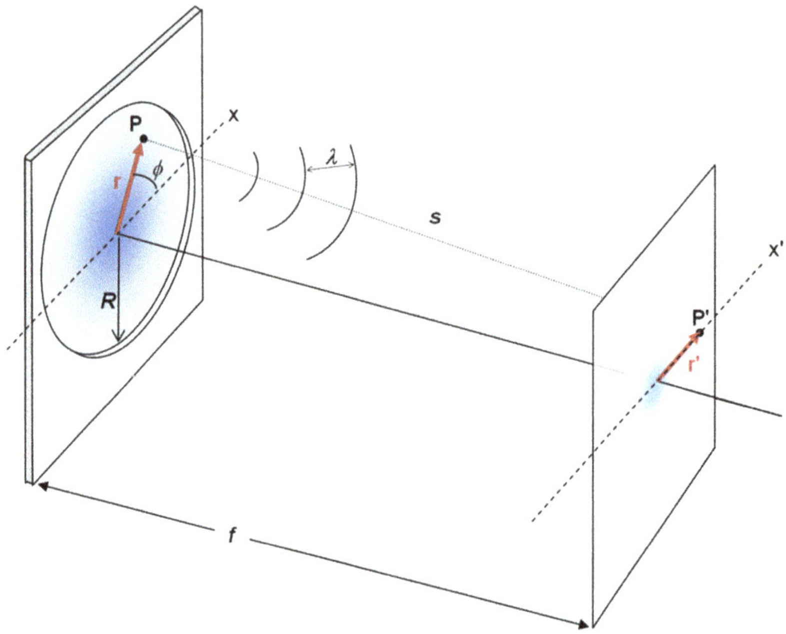

Figure 4.17. Schematic for focusing the pattern on the tube lens down to a real image.

objective, and the tube lens around an optical axis that defines the origins at each plane, point P′ can be chosen to be on an arbitrarily-chosen x′ axis.

The total field $E'(\mathbf{r}')$ at point P′ is due to the contributions of all the wavelets emanating from the electric field $E(\mathbf{r})$ (including phases) at the source plane, integrated over all $\mathbf{r}$. Then

$$E'(\mathbf{r}') = \int E(\mathbf{r})\frac{e^{iks}}{i\lambda s}d^2\mathbf{r}, \tag{4.39}$$

where

$$s^2 = f^2 + |\mathbf{r} - \mathbf{r}'|^2 = f^2\left[1 + \frac{|\mathbf{r} - \mathbf{r}'|^2}{f^2}\right]. \tag{4.40}$$

In an actual microscope, the relevant ranges of the three distances in equation (4.40) are all of different orders of magnitude: $|\mathbf{r}'| \sim 0.001|\mathbf{r}|$ and $|\mathbf{r}| \sim 0.01f$. Therefore, we can make approximations by rewriting the square root of equation (4.40):

$$s \approx f\left(1 + \frac{|\mathbf{r} - \mathbf{r}'|^2}{2f^2}\right). \tag{4.41}$$

Multiplying the squared term out, and eliminating terms that are second-order in r'/f, we get

$$s \approx f\left(1 + \frac{r^2 - 2\mathbf{r}\cdot\mathbf{r}'}{2f^2}\right). \tag{4.42}$$

Substituting this expression for s back into equation (4.39) gives

$$E'(\mathbf{r}') = \frac{e^{ikf}}{i\lambda f}\int E(\mathbf{r})e^{\frac{ikr^2}{2f}}e^{-\frac{ik\mathbf{r}\cdot\mathbf{r}'}{f}}d^2\mathbf{r}. \tag{4.43}$$

The field just after the tube lens, $E(\mathbf{r})$ in equation (4.37), can be substituted in, yielding:

$$E'(\mathbf{r}') = \frac{\alpha}{\lambda f}\int E_0(\mathbf{r})e^{-ik\mathbf{r}\cdot\mathbf{r}'/f}d^2\mathbf{r}. \tag{4.44}$$

Apart from the $(\lambda f)^{-1}$ scaling and phase factor α (which has unit amplitude) in front of the integral sign, equation (4.44) is simply a 2D Fourier transform. In other words, the electric field pattern at the focus is the Fourier transform of the electric field pattern at the lens. In the special case in which the illumination of the tube lens is uniform up to its radius R, equation (4.44) becomes:

$$E'(r') = \frac{\alpha E_0}{\lambda f}\int_{r=0}^{R} r dr \int_{\phi=0}^{2\pi} e^{-i\left(\frac{krr'}{f}\right)\cos\phi}d\phi. \tag{4.45}$$

Note that, in general,

$$J_0(x) = \frac{1}{2\pi}\int_0^{2\pi} e^{ix\cos\phi}d\phi \text{ and } \int_0^R xJ_0(x)dx = RJ_1(R), \tag{4.46}$$

where J_i is a Bessel function of order i. Therefore, the ϕ integral of equation (4.45) is equal to $2\pi J_0(krr'/f)$ As indicated, the subsequent r-integration produces a first-order Bessel function and the following result for the Airy PSF electric field:

$$E'(\mathbf{r}') = \alpha E_0\frac{R}{r'}J_1\left(\frac{kR}{f}r'\right). \tag{4.47}$$

As r' approaches zero (i.e. toward the optical axis in the image plane), the first-order Bessel function J_1 in the numerator approaches zero. However, the denominator also approaches zero, and the ratio approaches one, much like the behavior of $(\sin x)/x$ in one dimension. Figure 4.18 shows graphs of the first three orders of Bessel functions.

The Airy disk pattern is an example of a well-known result from Fourier optics: that the electric field at the focus of a lens is the Fourier transform of the electric field at the lens itself, as possibly truncated by the edge of the lens. The width of a Bessel function pattern is usually measured out to the position of the first zero crossing.

The intensity of the Airy PSF is:

$$I(r') = |E'(r')|^2 = E_0^2\left(\frac{R}{r'}\right)^2\left[J_1\left(\frac{kR}{f}r'\right)\right]^2. \tag{4.48}$$

Equation (4.48) is of interest only if the distance on the image plane r' is converted to an equivalent distance r' on the object. Referring to figure 4.17, $R/f \sim \theta_{\text{tube}}$.

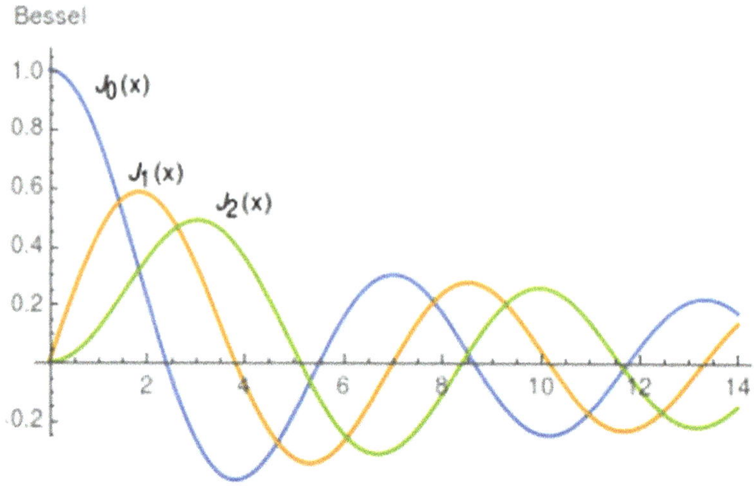

Figure 4.18. Graph of the three lowest-order Bessel functions. Reprinted with permission from [https://www.cfm.brown.edu/people/dobrush/am34/Mathematica/ch7/bessel.html].

The magnification of the whole system is $m = \theta_{obj}/\theta_{tube}$ so that $r' = mr'$. Therefore, the argument of J_1 becomes

$$\frac{kR}{f}r' = k\theta_{tube}\frac{\theta_{obj}}{\theta_{tube}}r'' = \frac{2\pi\theta_{obj}}{\lambda}r''. \tag{4.49}$$

J_1 has its first zero when the argument has the value 3.83, so at that equivalent radius (referred to the sample plane),

$$r''_{firstzero} = \frac{3.83\lambda}{2\pi\theta_{obj}} = \frac{0.61\lambda}{\theta_{obj}}. \tag{4.50}$$

This result was derived in the special case for which the angles are small and the refractive index on all sides of the two lenses is $n = 1$. A more complete calculation, which takes into account larger angles and the possibility that the objective is a liquid (water, glycerol, or oil) immersion model (with refractive index $n > 1$), would show the general answer to be

$$r''_{firstzero} = \frac{0.61\lambda}{n\ \sin\theta_{obj}} = \frac{0.61\lambda}{NA}, \tag{4.51}$$

where NA = $n\sin\theta_{obj}$ is the definition of the numerical aperture of the lens. This familiar formula, the standard 'Raleigh resolution limit,' is, however, only an approximation, since it still ignores polarization and the highly nonuniform pattern of fluorescence emission near an interface due to evanescent near-field emission capture by the coverslip. Nonetheless, it is convenient for use in deconvolving TIR images to improve sharpness, as will be discussed in section 6.4, and also in deriving expressions for spot TIR in section 5.11.

4.10.2 Full calculation of the PSF

To obtain a more accurate picture of the PSF that includes near-field evanescent effects and polarization, we use equation (4.34) along with digitally computed solutions for the factor in the integrand $\boldsymbol{E}'_{\mathrm{BFP}}$ (?? what is this compared to earlier work?). Figure 4.19 (panels a–i) shows the computer-calculated results for the PSF (an image of a single dipole) with three different orientations viewed through a linear polarizer (or not), based on the steps described above, going from the emission-field pattern (equation (4.18)) to the tube-lens focal-plane field (equation (4.35)) and then to I'_{IP}. As can be seen, the PSFs are much more complex that a simple symmetric blur. In the case of a dipole oriented at a 45° polar angle, the PSF is noticeably lopsided. If the azimuthal angle were $\varphi' = \pi$, the x-direction asymmetry would be reversed. However, if the supercritical light in the BFP is completely blocked (panel (k)), then the asymmetry disappears. The asymmetry arises entirely from the supercritical light gathered by the high-aperture objective. One practical consequence of this asymmetry is that determination of the dipole position (at the center)

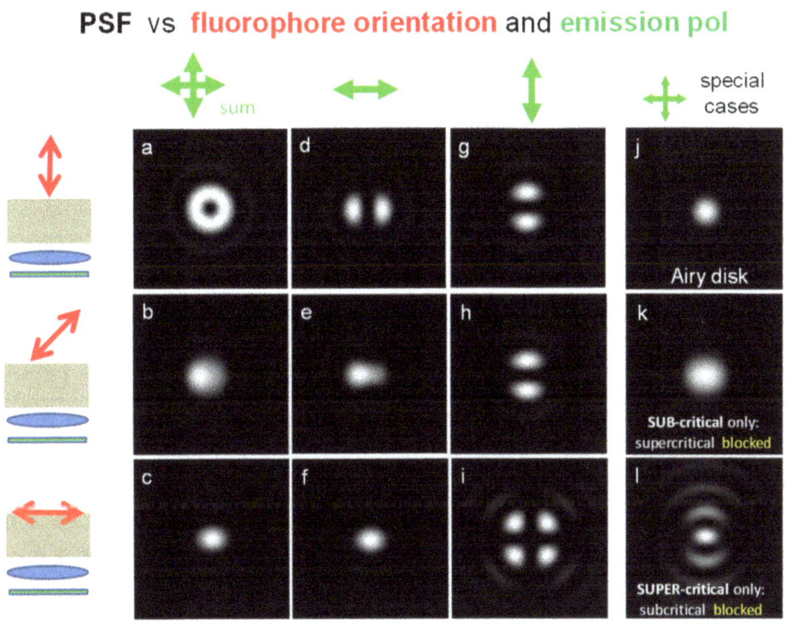

Figure 4.19. PSF intensity pattern at the image plane for the same cases as the BFP patterns in figure 4.9: $(\theta',\phi') = (0,0)$, $(\pi/4,0)$, and $(\pi/2,0)$, all with $h = 0$, $z' = 0$, NA = 1.49, $n_1 = 1.33$, and $n_3 = 1.52$. **(a–c)** NO emission path polarizer; **(d–f)** linear polarizer oriented to transmit x polarization only (left–right in the plane of this diagram); **(g–i)** linear polarizer oriented to transmit y polarization only (up–down in the plane of this diagram). Two special cases are also shown: **(j)** an Airy disk generated by completely filling the aperture with unpolarized light. The characteristic radius $w = 0.61\ \lambda$ /NA (= 0.409 λ for NA = 1.49) sets the scale for all the panels. **(k)** Supercritical BFP light completely blocked. **(l)** Supercritical only; subcritical BFP light completely blocked. The resulting PSF pattern, in this case for a dipole oriented left–right in the figure, becomes quite broad and has rings. Chapter 5 shows examples of how subcritical light can be excluded. Reprinted from Axelrod (2012), copyright (2012) John Wiley & Sons.

based on an experimental centroid-weighting procedure (as typically done in the photo-activated localization microscopy (PALM) and stochastic optical reconstruction microscopy (STORM) single-molecule tracking techniques, etc) gives inaccurate, laterally shifted results.

Panel (j) shows the standard Airy disk PSF, which is derived using the assumption that the dipole emission is completely isotropic, unpolarized, monochromatic, and of uniform phase across the entire objective aperture. This assumption is simple and is the basis for the Raleigh resolution condition, but in fact, it is actually never the case for a real dipole in proximity to a coverslip surface.

The PSF is also qualitatively altered in the presence of intermediate films of thickness h. Just as an example, figure 4.20 shows what happens to the PSF in figure 4.19 panel (a) as an intermediate film is introduced.

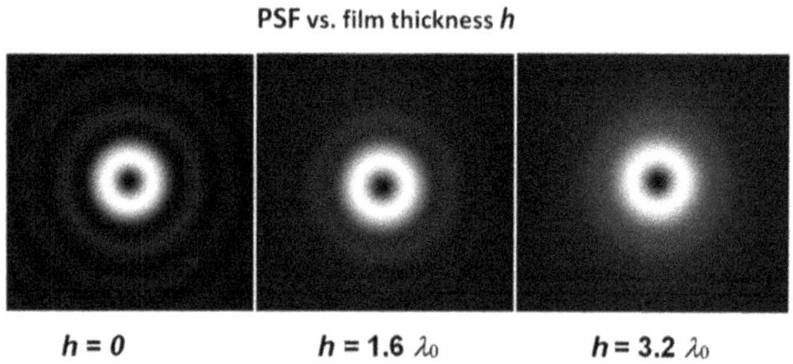

Figure 4.20. Effect of a dielectric film $n_2 = 1.42$ interposed between a perpendicularly oriented dipole and the surface. The dipole resides in n_1 at the n_1:n_2 interface. The parameters are the same as for figure 4.19. These PSFs result from the corresponding BFP patterns shown in figure 4.11.

4.11 Virtual supercritical angle fluorescence microscopy (vSAF)

Imaging of just the supercritical BFP annulus should select only those fluorophores that are close enough to the surface for the imaging to capture their near fields. This modification of a microscope is called 'supercritical angle fluorescence' (SAF) microscopy. It is easily accomplished by placing a circular opaque disk with an appropriate diameter at any place in the emission optics that is conjugate to the BFP. Figure 4.21 shows an actual example where this blockage has been employed. The system does produce an image of exclusively surface-proximal fluorophores, even with standard non-TIR (i.e. EPI) illumination. However, the images obtained are not crisp, as might be expected from the non-compact PSF shown in figure 4.19 (panel (l)).

To circumvent the poor lateral resolution of SAF microscopy, a modification has been introduced called 'virtual supercritical angle fluorescence' (vSAF). Instead of directly imaging just the light from the supercritical zone (by physically blocking the subcritical zone in the BFP), a whole-aperture (super- and subcritical) image is taken, and from it is subtracted the image from the subcritical zone alone (obtained by physically blocking the supercritical zone with a limiting aperture in the BFP or any of its conjugate planes). The difference image is a 'virtual' version of the direct

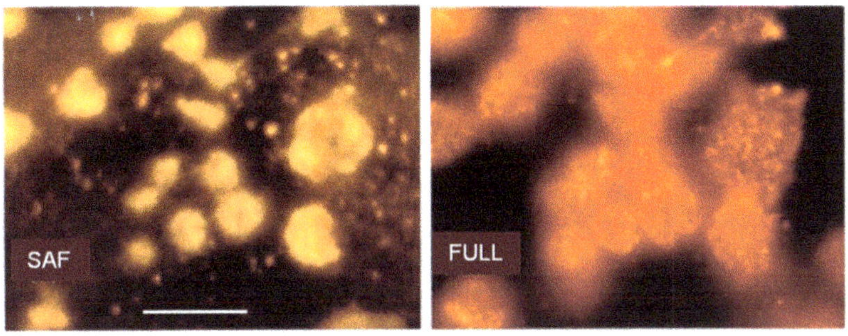

Figure 4.21. DiI-labeled bovine chromaffin cells; illuminated by EPI, NA = 1.65, 100X objective. The right-hand side shows an image obtained using the full aperture of the objective. The left-hand side shows an image obtained while the subcritical light was entirely blocked by an opaque disk at the BFP. (scale bar = 30 μm). Reprinted with permission from Axelrod (2012). CC-BY 4.0.

supercritical image, but it is better in one important respect: the difference image preserves the coherent electric field interference between super- and subcritical light coming from each fluorophore in the sample. Interference between super- and subcritical light does not occur at the BFP (because they in separated zones there), but it does occur at the image plane, where light from the two zones from each fluorophore on the sample recombines. Preserving that information-containing interference in vSAF preserves the lateral resolution, whereas losing that interference in direct supercritical imaging decrements the lateral resolution.

Theoretical reconstruction of the PSF for vSAF by the methods described above results in a much sharper PSF (figure 4.22) that still retains selectivity for surface-proximal fluorophores.

By combining the vSAF emission image protocol with standard TIRF excitation, an even higher degree of surface selectivity should be attainable than from either individually. There is an additional advantage to the combination. TIRF by itself has scattering artifacts, in part arising from the TIR evanescent field scattering at a heterogeneous sample (such as a layer of cells). This scattered light excites fluorophores farther away from the substrate than would normally be excited by a pure evanescent field. However, in combination with vSAF, these more distant fluorophores are discriminated against, probably resulting in a significant reduction of scattering background intensity.

4.12 Emission polarization including supercritical light

The ability of a high-aperture objective to see an oriented fluorophore from a continuous range of angles tends to reduce the observed polarization of the emission, even if the near-field effects covered in this chapter are neglected. However, if we take near fields into account, the depolarization effect becomes even stronger. Consider an excited single-molecule dipole residing in water immediately on a planar glass substrate, and oriented parallel to the surface, say in the x direction; the emission intensity F is observed through a linear polarizer oriented to transmit either in the x or y direction. A low-aperture objective should see emission light almost

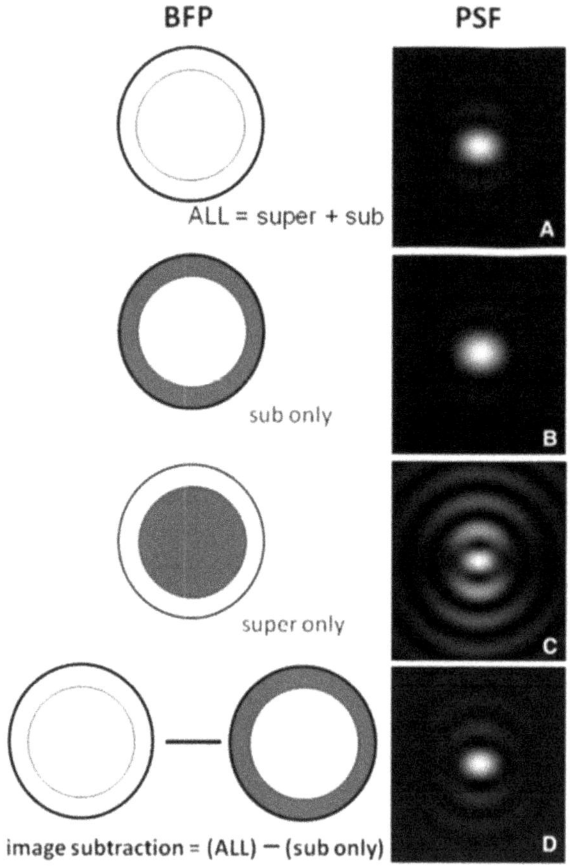

Figure 4.22. vSAF. The left-hand column shows a schematic of the BFP with (A) no blockage; (B) only the supercritical zone physically blocked; and (C) only the subcritical zone physically blocked. The right-hand column shows the theoretical PSF for each case, assuming a dipole oriented along x (left–right here) and a 1.49 NA objective. Although surface-selective, the super-only PSF of panel C has rather bright large rings that decrement the image quality; these are much reduced by the image subtraction scheme shown in panel D. Reprinted from Axelrod (2013), copyright (2013), with permission from Elsevier.

entirely polarized in the x direction; i.e. $F_y/F_x \sim 0$, and this is accurately predicted by calculations that both include and neglect the near field. However, for a 1.49 NA oil-immersion objective that can gather a considerable amount of near-field light through the supercritical zone of the BFP, a calculation that neglects the near field yields $F_y/F_x = 0.0394$, whereas a calculation that includes the near field gives $F_y/F_x = 0.135$.

4.13 SAF/UAF: measurement of the absolute distance between a fluorophore and a surface

Because the light passing through the supercritical zone of the BFP originates entirely from the fluorophore near field, the farther the fluorophore resides from the interface, the dimmer the supercritical light. This suggests a direct experimental method to determine the absolute z position of a fluorophore from the substrate: take the ratio of the integrated supercritical intensity to the integrated subcritical intensity from direct images of the BFP. Since 'subcritical angle fluorescence' unfortunately has the same acronym as 'supercritical angle fluorescence,' it is usually called 'undercritical angle fluorescence' (UAF), and so the ratio is called SAF/UAF. For a 1.49 NA objective and a dipole oriented parallel to the surface ($\theta' = 0$), the ratio starts at ~1.4 for a fluorophore immediately on the surface ($z'/\lambda_o = 0.0$) but rapidly drops to ~0.85 at $z'/\lambda_o = 0.1$ (see figure 4.23). The ratio continues to decrease to 0.06 at $z'/\lambda_o = 1.0$. Given any particular supercritical-to-subcritical ratio, the corresponding absolute z can be read. The key advantage of using this ratio to estimate z is that it depends upon neither calibration of the observed intensity nor normalization by the total emitted power: the ratio is inherently normalized against changes in illumination power and emission and collection efficiencies.

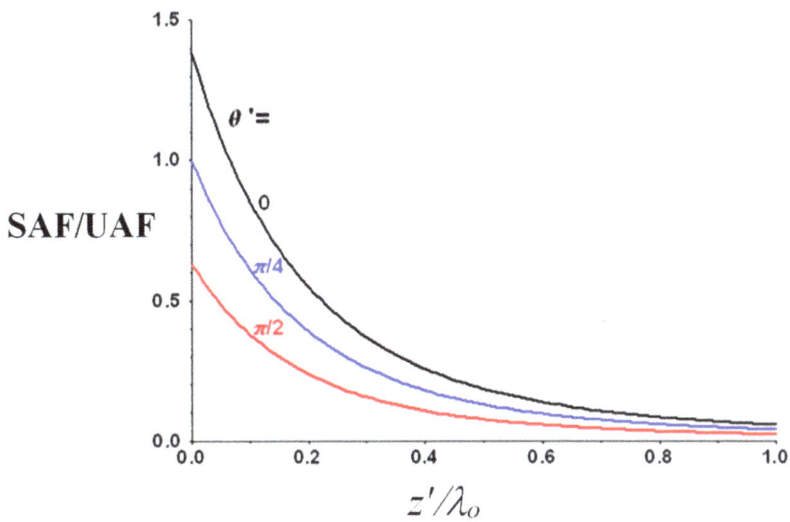

Figure 4.23. SAF/UAF ratio vs z'/λ_o. The parameters assumed are $h = 0$, $z' = 0$, NA = 1.49, $n_1 = 1.33$, and $n_3 = 1.52$. Reprinted from Axelrod (2014), copyright (2014), with permission from Elsevier.

Experimentally, the SAF/UAF technique requires that extra care be taken to avoid (or to account for) any background light (from the room or from scattering by the sample itself), because such a background will appear largely in the UAF zone and thereby disrupt the ratio.

4.14 Effect of near-field capture on fluorescence lifetime

The larger the value of P_T, the more rapidly the dipole loses its excitation energy into its surroundings (as light or heat). Therefore, making the conceptual jump to viewing the dipole as a quantum system, it is intuitively clear that as P_T increases, the fluorescence lifetime should decrease. Here, we discuss general expressions for the quantum efficiency q and lifetime τ of a fluorophore (modeled as a fixed-power dipole) near an interface. In contrast to the previous sections, in which we assumed that the quantum efficiency of the isolated ($z' << \lambda_o$) fluorophore was unity, here, we assign it a more general value, q_0, which may be less than unity (possibly because of quenching arising from the surrounding chemistry). The following definitions are useful:

$$p(z') \equiv \frac{P_T(z')}{P_T(z' = \infty)}$$
$$f(\Omega', z') \equiv \frac{\int d\Omega S(\Omega, \Omega', z')}{\int d\Omega S(\Omega, \Omega', z' = \infty)}, \tag{4.52}$$

where the $\Omega = (\theta, \phi)$ integration is over all 4π steradians, $\Omega' = (\theta', \phi')$ is a shorthand notation for the dipole orientation, and the calculations of P_T and S are as described in equations (4.18–4.23). The quotient p is the ratio of the total dissipated (light plus heat) power of a fixed-amplitude dipole located at a distance z' from a surface to the total power at infinity. The quotient f is the ratio of just the radiated (i.e. light) portion of the power released by a fixed-amplitude dipole positioned at at distance z' to the radiated power of the same dipole at infinity.

Let q be the orientation-dependent quantum efficiency of the fluorophore at distance z'. One can show that

$$\frac{1}{q} = \frac{\frac{1}{q_0} + p - 1}{f}. \tag{4.53}$$

Given a natural fluorescence lifetime τ_0 for a fluorophore at z' = infinity in the absence of any radiationless decay pathways, one can show that the actual observed lifetime τ for a real fluorophore near an interface will be

$$\tau = \frac{q\tau_0}{f}. \tag{4.54}$$

This physical effect of dielectric surface proximity on fluorescence lifetime is distinct from any chemical effects of surface proximity.

Further reading

Axelrod D 2012 Fluorescence excitation and imaging of single molecules near dielectric-coated and bare surfaces: a theoretical study *J. Microsc.* **247** 147–60

Axelrod D 2012 Selective imaging of surface fluorescence with very high aperture microscope objectives 6 6–13 *J. Biomed. Opt.* **6** 6–13

Axelrod D 2013 Evanescent excitation and emission in fluorescence microscopy (Invited review) *Biophys. J.* **104** 1401–9

Axelrod D 2014 Evanescent excitation and emission *Fluorescence Microscopy: Super-Resolution and other Novel Techniques*; A Cornea and P M Conn (Amsterdam: Elsevier) ch 1 pp 1–14

Born E and Wolf E 1999 *Principles of Optics: Electromagnetic Theory of Propagation, Interference and Diffraction of Light* 7th edn (Cambridge: Cambridge University Press) ch 1

Enderlein J, Gregor I and Ruckstuhl T 2011 Imaging properties of supercritical angle fluorescence optics *Opt. Express* **19** 8011–8

Ford G W and Weber W H 1984 Electromagnetic interactions of molecules with metal surfaces *Phys. Rep. (Rev. Sec. Phys. Lett.)* **113** 195–287

Hellen E H and Axelrod D 1987 Fluorescence emission at dielectric and metal-film interfaces *J. Opt. Soc. Am.* B **4** 337–50

Chapter 5

Optical configurations and setup

A wide range of optical configurations can be employed for total internal reflection fluorescence (TIRF) microscopy. In general, an inverted microscope is more convenient because it (a) provides more room to add total internal reflection (TIR) optics above rather than below the stage, and (b) accommodates glass-bottomed open cell culture dishes. However, some upright microscope configurations are still very workable and have unique features.

Many configurations use a prism ('prism-based-TIRF') to direct light toward the TIR interface—and this has some advantages—but most use the microscope objective itself for this purpose ('objective-based-TIRF'). The main commercial microscope manufacturers all sell premade objective-based TIRF systems.

Commercial systems generally work 'out of the box.' Among other topics, this chapter explains how commercial objective-based TIRF works. However, there are many reasons for custom designing and building a system: a custom-built system is more amenable to modification for special variations, and it is generally much cheaper than a commercial system. This chapter also covers in detail how to build and align custom-based systems, both objective and prism-based, with particular emphasis on those configurations appropriate for studies of cell cultures or adsorbed molecules on a glass coverslip. Immediately after the sections on prism-based TIR and objective-based TIR, hints are presented to help with the alignment procedure.

In all cases, the critical angle must be considered in the design. The refractive index of the standard glass coverslip upon which cells are grown is about $n_1 = 1.52$. The refractive index of the intact cell interior can be as high as $n_3 = 1.38$. Therefore, to obtain TIR at this interface, the angle of incidence must be larger than the critical angle of 65°. If the cells are not intact (e.g. if they are permeabilized, hemolyzed, or fixed) so that the lower refractive index is that of the aqueous buffer ($n_1 = 1.33$) instead of cytoplasm, then the critical incidence angle is 61°.

doi:10.1088/978-0-7503-3351-1ch5

In general, objective-based TIRF has the advantage of requiring no additional optics in the vicinity of the sample; all that is needed there is a very high-aperture objective. However, since the excitation beam passes through the objective, it generates some scattering from multiple reflections between objective lens elements, and some (out-of-focus) fluorescence excited in the objective glass itself. In addition, the range of incidence angles is limited by the numerical aperture of the objective, and the area of illumination is considerably smaller (typically <50%) of the full field of view.

Prism-based TIRF has much less excitation scatter and fluorescence background, a potentially much larger area of illumination, and, in some configurations, a wider range of incidence angles. When objective-based illumination is used, the full power of the illumination light excites some fluorescence in the glass and glues in the objective. In prism-based TIR, this background is eliminated, thereby making fluorescence detection of single molecules more feasible. However, the sample may be less accessible, the range of usable objectives may be more limited, and the optical systems need to be constructed and aligned on a do-it-yourself basis.

5.1 Inverted microscope TIR with prism above

Figure 5.1 shows several schematic drawings designated A–D for setting up TIR in an inverted microscope with a prism over the sample, all using a laser (usually with blue or green output) as a light source.

In all these configurations, the buffer-filled sample chamber consists of a lower bare glass coverslip, a spacer ring (often made of 60 μm thick Teflon) and the cell coverslip

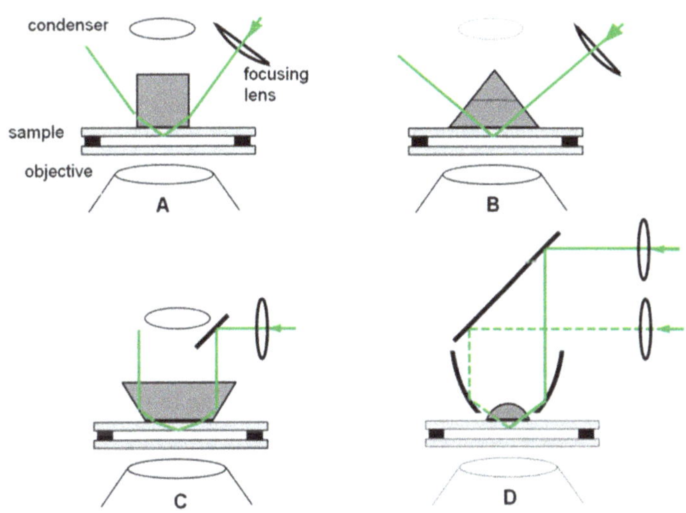

Figure 5.1. Four configurations (A–D) for an inverted microscope with a TIR prism above the sample. The sample is a coverslip sandwich with an intervening space; all vertical distances are exaggerated for clarity. In configuration B, the dotted lines refer to the option of truncating the prism in order to clear the optical path for a condenser system. In configuration D, the option of producing interference fringes using intersecting beams is depicted by a dashed line. Reproduced from Axelrod (1999), copyright (1999) Oxford University Press.

inverted so that the cells face down. The upper surface of the cell coverslip is put into optical contact with the prism lowered from above by a layer of immersion oil or glycerol. The lateral position of the prism is fixed, but the sample can be translated while still maintaining optical contact. The lower coverslip can be oversized and the Teflon spacer can be cut with gaps so that solutions can be changed by capillary action with entrance and exit ports. Alternatively, commercial solution-changing chambers or rectangular cross-section microcapillary tubes can be used.

The configurations in this set share some common advantages:

- inexpensive optics
- ample room for setup
- the prism may be mounted on a condenser holder for ease of raising and lowering
- the ease of checking alignment

However, these configurations share some common drawbacks:

- the sample is not easily accessible from above
- shortest working-distance objectives may not achieve focus across the buffer layer
- the image quality with the highest-aperture objectives is somewhat reduced by viewing through the buffer layer

The four configurations A–D each have their own separate advantages and disadvantages:

Configuration A: The rectangular solid prism allows transmitted light from the microscope condenser above to reach the sample undistorted, so that phase contrast and dark field can be viewed simultaneously. Such rectangular solid prisms must be custom cut and polished, but a commercial polarizing cube (PC) can be substituted for the prism. The size of the TIR illuminated region and the incidence angle are easily varied by adjusting the position and angle of the incoming laser beam and the focusing lens.

Configuration B: The triangular prism is commercially available in numerous sizes and is inexpensive; it can generally be obtained in 45–45–90 right-angled and 60–60–60 equilateral shapes. The maximum incidence angle is obtained by introducing the beam from the horizontal direction. For standard glass ($n = 1.52$), the maximum incidence angle is 73° for the right-angle prism and 79° for the equilateral prism. Phase contrast and dark field are not compatible with this configuration, because the upper surface of the triangular prism is not flat. However, custom truncation and polishing of the top of the prism (shown as an option in figure 5.1B) provide compatibility with phase contrast and dark field.

Configuration C: The 60° trapezoidal prism is the most convenient and reproducible system of all when mounted on the microscope condenser mount. The incoming beam is vertical, so the TIR spot shifts laterally by very little when the prism is raised and re-lowered during changes of sample. Phase contrast and dark field are compatible with this flat-topped prism. The incidence angle is fixed at 60°.

As discussed above, this angle is not sufficient to produce TIR if the prism is made from standard $n_3 = 1.52$ glass. Therefore, the prism *must* be made of higher-than-standard refractive index material (e.g. flint glass at $n = 1.64$). The beam will then refract away from the normal to $\theta = 69°$ in passing from the prism into the coverslip, thereby exceeding the critical angle at the coverslip/water or coverslip/cell interface. Ideally, a trapezoid with walls between 45° and 60° would be best, but this might have to be fabricated as a more expensive special order. Neither 45° or 60° trapezoids are commercially available either, but they can be cheaply manufactured by truncating and polishing the apex of a commercially available triangular prism.

Configuration D: The parabolic mirror and hemispherical prism here are positioned so that the beam traverses a radius of the prism toward a TIR spot at the focus of the parabola. In this manner, a lateral shift of the vertical incoming beam will always focus at the same spot. Substantial changes of incident angle can thereby be accessed quite conveniently. In addition, interference fringes can be created in the TIR evanescent field by splitting the incoming beam into two beams, each reflecting at different azimuthal positions in the parabola but recombining at the same parabola focus. The spacing of the fringes can be adjusted by varying the relative azimuthal positions of the two beams. Despite all this versatility, this setup is fairly difficult to align and it can be rather sensitive to vibrations.

Configuration E: For some experiments, it is desirable to change solutions rapidly and follow the TIRF vs. time of small objects (e.g. cells) in response to the chemical jump. Figure 5.2 shows an inverted microscope TIRF setup in which the prism

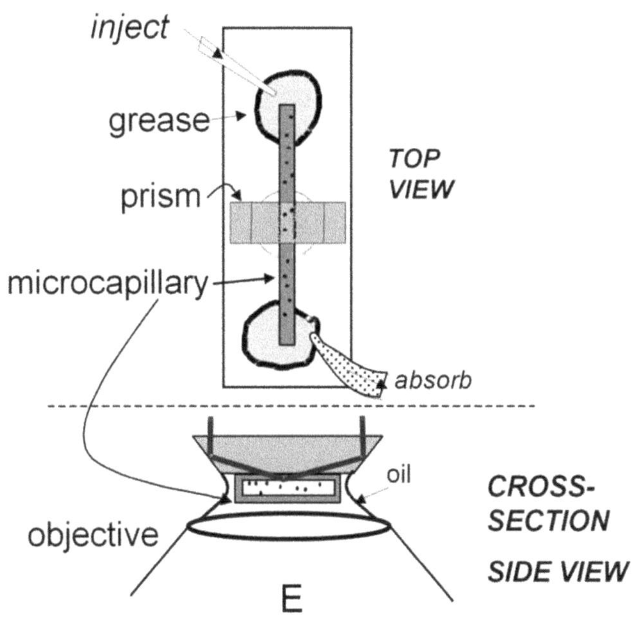

Figure 5.2. Configuration E: prism-based TIRF in contact with a microcapillary tube for quick solution changes. Reprinted from Axelrod (2003), copyright (2003) with permission from Elsevier.

above the sample is in contact with a rectangular cross-section microcapillary, through which rapid solution changes are possible.

5.2 Inverted microscope TIR with prism below

TIR can be set up to be compatible with simultaneous microinjection or micro-electrophysiology. To provide for easy and continuous access to a liquid-containing sample from above, a prism TIR setup should have the prism deployed below stage level. Unfortunately, that is also where the objective resides, so the geometry is rather tight. Configuration F (shown in figure 5.3) accomplishes this goal by taking advantage of multiple TIR in the coverslip. This functions as a waveguide, transferring the excitation light from a far off-axis to the center of the field of view.

In the simplest form of this configuration, only a small commercially available triangular prism need be placed in optical contact (via oil) with the bottom of the cell-containing coverslip. The sample can be translated while the prism remains laterally fixed. The problem is that this may leave a smear of oil on the lower side of the coverslip, which destroys the first internal reflection. A very feasible alternative is to use an additional intervening coverslip fixed to the prism with optical glue. The sliding motion then occurs between the intervening coverslip and the cell coverslip, which are lubricated and optically contacted by a thin layer of immersion oil.

The disadvantage of this setup is that oil- or glycerol-immersion objectives cannot be used, because this might destroy the TIR of the final internal reflection before the spot under view. However, water (or air) immersion objectives work well with this configuration.

Configuration F can be modified to allow the beam to enter the lower slide by means other than from below via a prism. Here are three possibilities. One modification is to extend the lower coverslip's area beyond the edge of the cell culture dish and place the prism on top of it, thereby giving it room to capture the beam from above. Another modification, conceptually similar to Configuration F, is to extend the lower coverslip similarly, but instead of a prism, use a blazed

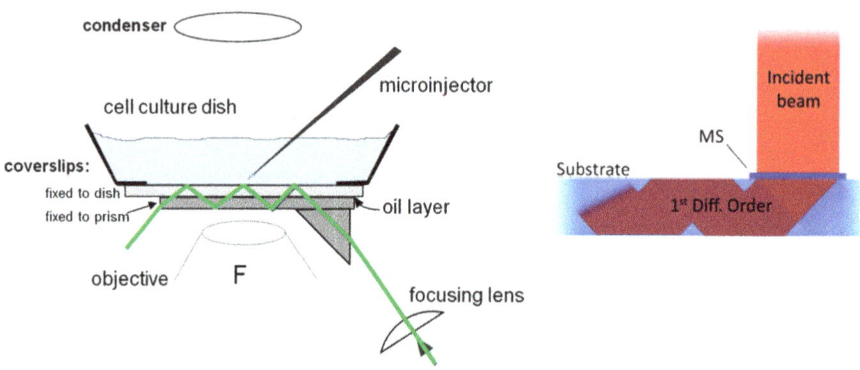

Figure 5.3. (Left panel) Configuration F for an inverted microscope with a TIR prism below the sample. The lower coverslip (shown darker) is fixed to the prism using optical glue. Reproduced from Axelrod (1999), copyright (1999) Oxford University Press. This coverslip is optional, intended to avoid smearing immersion oil. (Right panel) The 2D metasurface (MS) creates supercritical first-order diffraction. Reprinted from Gortari *et al* (2020), copyright (2020) Optical Society of America under the terms of the OSA Open Access Publishing Agreement.

diffraction grating placed flat and in optical contact with the top of the coverslip. The diffraction grating blazing would be designed to send most of the light (impinging normally from above) into a first-order beam of sufficient angle to achieve TIR. This method is illustrated in the right-hand panel of figure 5.3, using a specially designed and fabricated '2D metasurface' (MS), essentially a high-efficiency diffraction grating. A third modification is to use a slightly thicker lower coverslip with an optically polished edge, so that it can directly accept a supercritical beam.

The disadvantage of configuration F and its variations is that the sample itself presents refractive-index nonuniformities that tend to impair the uniformity of the incident beam. The degradation is cumulative with each reflection, so only the first two or three reflections might be usable.

5.3 Upright microscope TIR with prism below

Possibly the most convenient prism-based TIR setups of all are configurations G and H, shown in figure 5.4. Configuration G can use the same kind of trapezoidal prism as that discussed in section 5.1 above, mounted on the microscope's condenser holder. Alternatively, a hemispherical prism can be used. The laser beam is introduced into the same port in the microscope base as that intended for the transmitted light illuminator (which should be removed), thereby utilizing the microscope's own in-base optics to direct the beam vertically upward. An extra lens just above the microscope base may be useful to position and focus the TIR spot. As discussed earlier, flexibility in incidence angle is sacrificed for convenience; however, a set of various-angled trapezoids will allow one to employ various discrete incidence angles.

This system gives particularly high-quality images if a water immersion objective is employed and submerged directly into the buffer solution in an uncovered cell chamber. This system is also easily used with cells adhering directly to tissue culture plastic dishes rather than on coverslips; the plastic/cell interface is then the site of TIR. If the objective has a long enough working distance, reasonable accessibility to micropipets is possible.

Configuration H is not strictly 'prism'-based, but the function is the same. A high-aperture condenser (oil immersion) is used instead of a prism. For light to emerge as

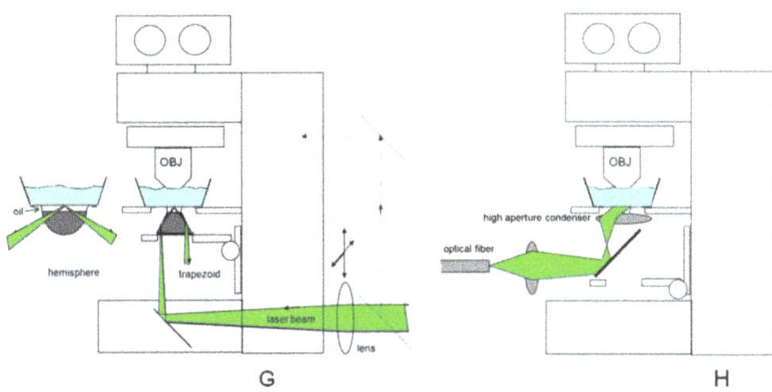

Figure 5.4. Configuration G for an upright microscope with a trapezoidal or hemispherical prism below the sample. Reprinted from Axelrod (2001), copyright (2001) John Wiley & Sons. Configuration H uses a high-aperture condenser lens in place of the prism.Reprinted from Axelrod (2007), copyright (2007) with permission of Springer.

a collimated beam (so that all 'rays' are traveling at the same incidence angle), the beam must be focused at the condenser's back focal plane (BFP).

5.4 Objective-based TIR

Microscope objectives are commercially available with numerical apertures (NA) up to 1.5 (with standard immersion oil) or up to 1.7 (with specialized high-index oil). Using an objective with a numerical aperture of at least 1.4 or higher, supercritical-angle incident light can be cast upon the sample by illumination through the objective.

5.4.1 Focus at the back focal plane (BFP)

The incident beam must be constrained to pass through the periphery of the objective's pupil and must emerge with only a narrow spread of angles. This can be assured by focusing the illumination beam off-axis at the objective's BFP (figure 5.5). (The requirement for focus can be somewhat relaxed if a wider range of emergent angles can be tolerated; see section 5.11.1.)

The beam emerges into the immersion oil (refractive index n_3) at an angle θ, which must be less a maximum angle θ_{max} determined by:

$$NA = n_3 \sin \theta_{max}, \tag{5.1}$$

where NA is the numerical aperture of the objective. The incidence angle at the upper surface of the coverslip (the TIR surface), shown in purple in figure 5.5 differs from θ if the refractive index of the coverslip is different from that of the oil. But only the product $n \sin \theta$ matters in this discussion, and that product is constant for a beam

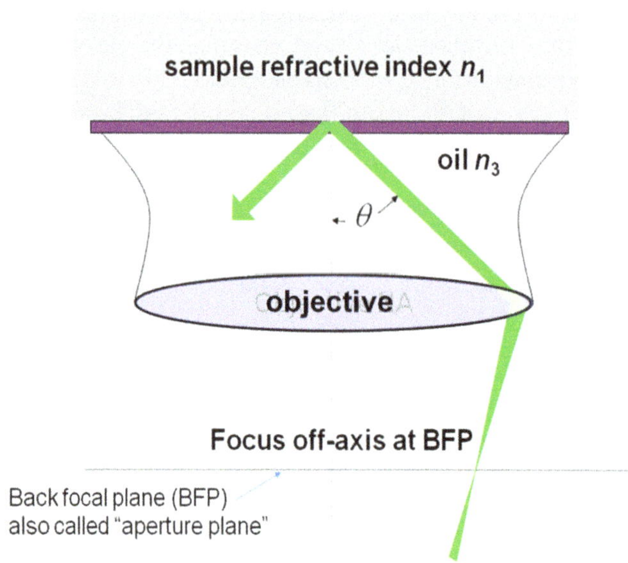

Figure 5.5. Objective-based TIRF with illumination beam focus at the BFP.

passing through any planar interface, according to Snell's Law. In addition, the refractive indices of the oil and the coverslip are usually quite similar.

For total internal reflection to take place at the sample surface, θ must be greater than the critical angle θ_c given by

$$n_1 = n_3 \sin \theta_c. \quad (5.2)$$

From equations (5.1) and (5.2), it is evident that for TIR to occur, it must be the case that

$$\mathrm{NA} = n_3 \sin \theta_{\max} > n_3 \sin \theta_c = n_1 \quad (5.3)$$

and therefore

$$\mathrm{NA} > n_1. \quad (5.4)$$

This requirement is no problem for an interface with water (for which $n_1 = 1.33$) and an NA = 1.4 objective. But in order to view the inside of a cell at $n_1 = 1.38$, this configuration is marginal at best for producing TIR, and higher-NA objectives are better.

The farther off-axis the focused beam at the BFP, the greater the incidence angle (and the thinner the evanescent field). The angle of convergence of the beam at the BFP (i.e. how steeply the rays converge) does not affect the incidence angle. However, that angle of convergence completely determines how wide a field is illuminated on the same (see the next subsection). Usually, the goal is to illuminate a broad region, or at least one as broad as the microscope camera's field of view. The wider the angle of convergence, the wider the illuminated field. This is why the laser beam should be expanded (by lens L1 shown in figure 5.6) to a wide area on a

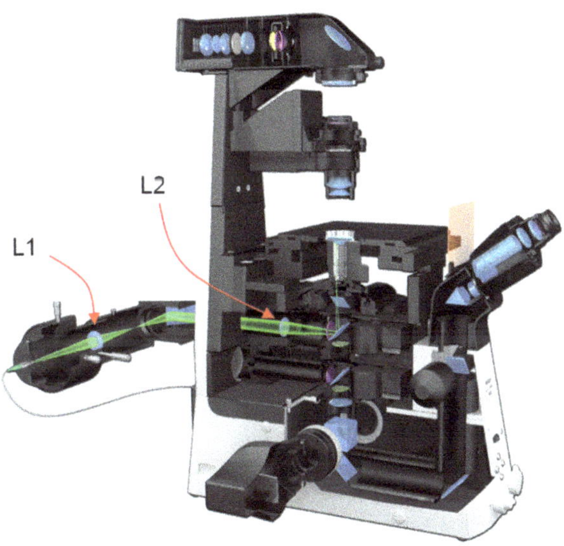

Figure 5.6. The optics in a simple commercial TIRF system. Courtesy of Nikon Instruments.

BFP-focusing lens (shown as L2), and why the BFP-focusing lens should be as close to the BFP (i.e. should have as short a focal length) as possible, consistent with steric constraints.

The advantages of objective-based TIR in an inverted microscope include the possibility of viewing a sample that is completely accessible from the top, and the compatibility with (in fact, necessity for) the highest-aperture, highest-resolution, brightest objectives available. This arrangement is also highly compatible with the use of intersecting beams to produce interference fringes (discussed below): the high mechanical stability enables one to achieve interfringe spacings of ~0.3 microns without the blurring effects of small vibrations.

In its simplest forms (figure 5.7), objective-based TIR can be arranged in a variety of ways, depending on the location of the entrance port of the microscope's illuminator unit (side vs rear) and upon the light source type (laser vs arc). LED light sources are excellent too, because they are very small and bright. This means that a simple lens can collimate the LED output into a collimated beam that can be treated much as a laser beam.

A variation of figure 5.7D illuminates the sample from a complete azimuthal circle of directions. With incoherent sources (such as arc lamps or LEDs), the evanescent intensities produced by light arriving from different directions simply add at the sample without producing interference. The main advantage of this approach is to suppress comets and shadowing artifacts at refractive index inhomogeneities,

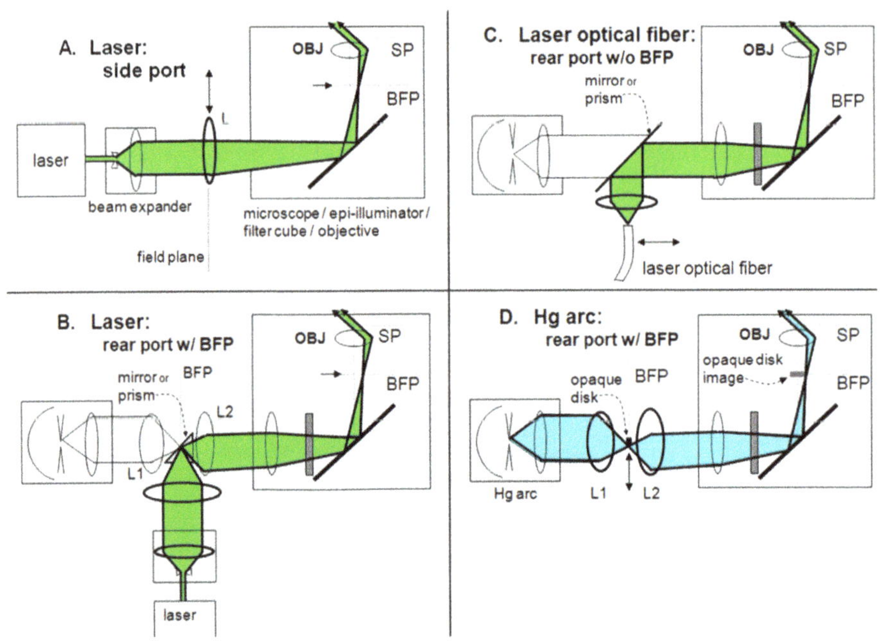

Figure 5.7. Simple objective-based schemes for introducing illumination into the fluorescence illuminator unit with the aim of achieving focus at the BFP. Reprinted from Axelrod (2001), copyright (2001) John Wiley & Sons.

which is similar to the simulation of that effect by azimuthal spinning of a single direction. Figure 5.8 shows an optical setup that uses a commercially available conical lens.

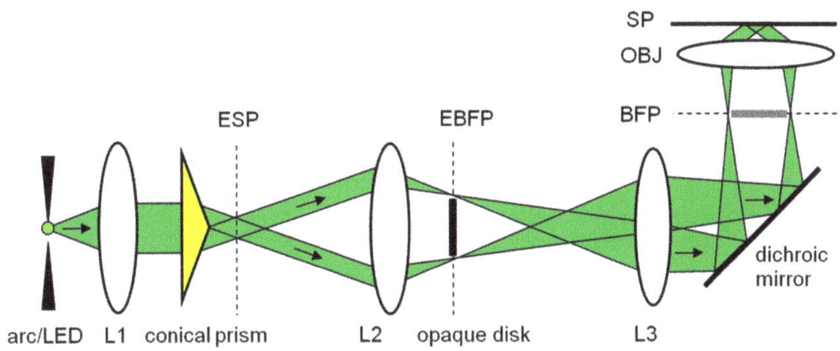

Figure 5.8. Optical setup used to produce a complete azimuthally isotropic ring of (unpolarized, incoherent) illumination at the BFP.

The result of this system is the production of a hollow cone of illumination with a constant-thickness wall and a vertex at the TIR/sample plane (SP) surface. The half angle of the cone is designed (using an appropriate choice of focal lengths and positions) to be greater than the TIR critical angle. The opaque disk is sized so that its shadow image at the BFP blocks all of the subcritical zone, but leaves much of the supercritical zone unblocked. The conical lens and lens L2, which convert a single wide collimated beam into a hollow cone and then to a sharp ring at the equivalent BFP (EBFP), serve to greatly increase the light intensity by bypassing the opaque disk and thereby appearing in the BFP supercritical zone. The conical lens and L2 are not strictly necessary, but without them, much of the light would be lost at the opaque disk.

5.4.2 Illumination area in the field of view

All of the above objective-based schemes produce a wide area of TIR illumination at the sample. That happens because the beam, in all cases, converges and focuses at the BFP, then diverges and is forced into collimation by the objective. Collimation ensures that all the rays approach the sample with the same incidence angle, so that the evanescent field has a single exponential decay depth. Clearly, the linear dimension of the illumination reaching the sample can be made smaller by reducing the angle of convergence/divergence on its way through the BFP: smaller angles do not expand as much before they are recollimated by the objective. The area of illumination can thereby be altered simply by adjusting the width of the collimated beam before it enters the whole system. This adjustment can be made without changing the positions or focal lengths of any downbeam optical elements. The width of a collimated beam can be set by a commercial or homemade 'beam expander.'

Later in this chapter (section 5.11), we will discuss objective-based systems that can produce structure in the TIR illumination pattern: spots, multiple spots, or stripes and the purposes for which such patterns can be used.

5.5 Incidence angle, multicolor, and polarization control

It is often desirable to have more versatile or even computer-controlled positioning of the focus at the BFP, so that incidence angle can be fine-tuned and changed quickly. This is particularly useful when comparing EPI with TIR (or near-TIR, in which the light propagates in a thin layer at a highly oblique angle in the sample). As shown in figure 5.9, optical control is achieved by a pair of small (commercially available) galvanometer mirrors ('galvos,' GM) or a pair of acousto-optic modulators (AOMs), one for each off-axis displacement of the focused beam at the BFP. The correct positioning of these beam-steering elements is crucial. The GM or AOM must be placed at a plane equivalent to the SP, so that when the GM is rotated, the illuminated area of the sample does not shift laterally.

Another desirable feature is the capability to switch from one excitation color to another to excite a different set of fluorophores. The challenge here is to make sure that both colors produce TIR at the sample, despite chromatic aberrations and dispersion in the optics. The lens L1 in figure 5.9, positioned only in the blue light path, compensates for the chromatic aberrations of lenses L2, L3, and the objective, all of which are located where the two paths coincide.

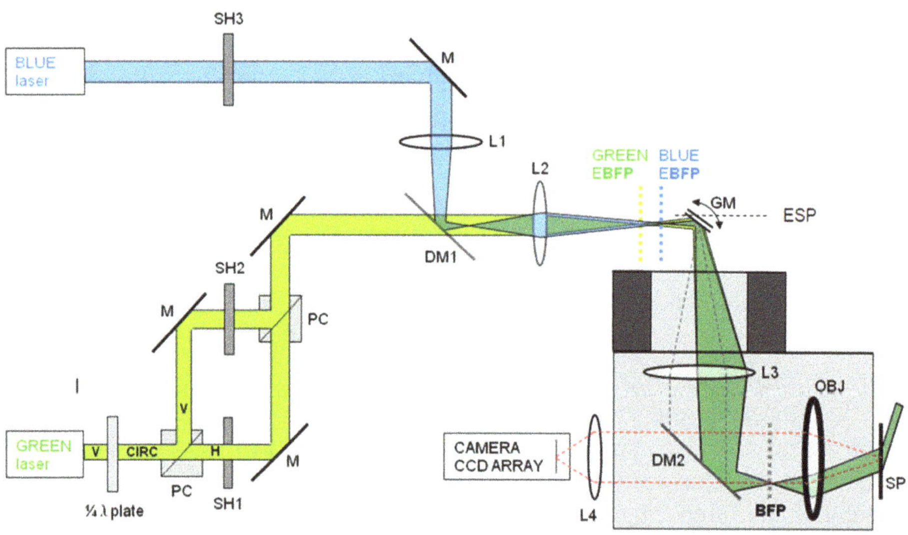

Figure 5.9. This is a view looking down from above at a custom TIRF optics setup through a microscope's rear port that combines manual control of the focus at the BFP, automated incidence angle control, and the capability to automate the switching of excitation color and polarization. At the illuminator's dichroic mirror (DM2), the drawing's viewing perspective changes so that the beam path can be depicted more clearly: the reflection from DM2 is actually up out of the page rather than to the right. Reprinted from Axelrod (2016), copyright (2016) with permission from Elsevier.

A third feature, to be discussed in detail in its own section, is the capability to switch rapidly between alternate excitation polarizations. This makes it possible for the system to detect fluorophore orientations.

The scheme shown in figure 5.9 builds all these features into one system.

5.5.1 Sample plane, back focal plane, and their equivalents

The central task in producing objective-based TIRF is to be sure that the beam is focused at the BFP, despite the fact that the exact location of the BFP (usually deep inside the objective) is neither known nor seen. Lens L3 is mounted on a fine-control (manual) translator; its job is to focus the beam at the BFP. This is judged by placing a flat coverslip in optical (oil) contact with the objective, focusing the microscope on its upper surface (made visible by sharp dust specks or a marker pen imprint) and viewing the size of the beam projected on the ceiling of the room. (Don't look into the beam!) Moving L3's translator will achieve a minimization of the diameter of the spot on the ceiling. This outcome necessarily means that the beam emerging from the objective is collimated, because the focus is correctly located at the BFP.

Lens L3 (typical focal length ~ 75 mm) also has another function: once it is emplaced, it casts an 'image' of the SP (an equivalent SP (ESP)) at a position immediately behind the microscope's rear illumination port. (Without L3, that image would be at infinity, since most modern microscope systems are designed to have an 'infinity space' behind the objective when the sample is in focus.) We want the GM set to be at that ESP. The laser beam remains at a fixed position at that ESP (presumably on the GM), even if the GMs are rotated. This ensures that as the GMs rotate, the illuminated region remains at a fixed position on the actual SP. Also, the larger the beam that is expanded on the ESP/GM, the larger the illuminated field on the actual SP.

Aside from relocating the ESP to immediately behind the microscope tower, L3 also acts to image the BFP at an EBFP. This EBFP position is somewhere upbeam from the GM/ESP. We want the laser beam to focus to a point at that EBFP, so that it refocuses to a point on the actual BFP. Lens L2 does that job, but it also has an additional function. By focusing the beam at the EBFP, it ensures that the beam is quite expanded by the time it gets to L3, thereby ensuring a high angle of convergence and a consequently large area of sample illumination. The longitudinal position of L2, in combination with the longitudinal position of L3, affects whether the beam is focused at the BFP. As an alternative to adjusting L3 to achieve the minimum-sized ceiling spot, L2's position can be adjusted with the same goal.

5.5.2 Polar incidence angle control

The angular positions of the two stacked GMs, which rotate around orthogonal axes, determine the location of the focused spot on the BFP without affecting the illumination region's position. Of course, two GMs cannot both be exactly at the ESP, but the ESP plane can be adjusted to be in the middle simply by slightly moving L3 longitudinally. As either of the GMs is rotated, the laser spot on the ceiling will head down the walls of the room, corresponding to a changing angle of incidence at the SP.

Note that the critical angle at which TIR begins is a function of the refractive index of the material above the TIR surface. Generally, water or saline will be used in an experiment. Here, we discuss a convenient method for confirming when TIR is achieved with the desired medium. It is best to replace the dry coverslip on the sample stage with a glass coverslip-bottomed cell culture dish, again put into optical contact (via immersion oil) with the objective. The coverslip can be pre-coated with an ethanolic solution of commercially available carbocyanine dye (such as diI-C_{18}-(3)). This fluorescent dye, which emits in the orange, has a positively charged head group and a hydrophobic tail, so it sticks to the glass in a monolayer and does not wash off with subsequent rinsing with water. Then fill the dish with a dilute (barely yellow) solution of aqueous fluorescein, which emits in the green, and again focus on the top of the coverslip surface. As the GM is rotated from subcritical ('EPI') to supercritical (TIR), the field of view seen through the microscope eyepiece will quickly transition from green (predominantly bulk excitation of fluorescein) to orange (almost exclusive excitation of diI). This effect can be seen in figure 5.10.

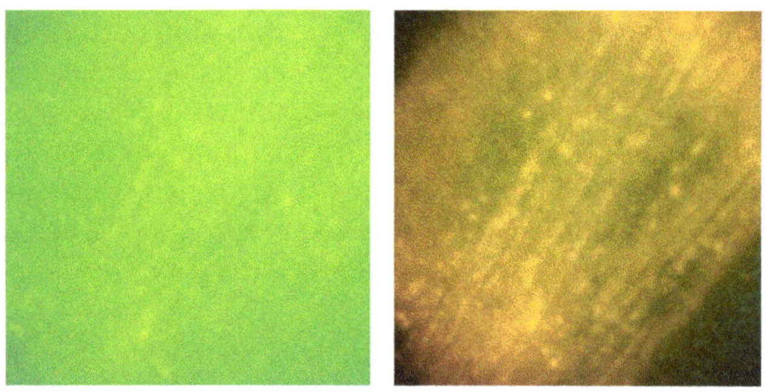

Figure 5.10. Transition from subcritical (left) to supercritical illumination (right) on a diI (on surface)/ fluorescein solution sample. The fluorescence in these images was excited by a mercury-arc source, as shown in figure 5.6(D), but the effect is similar with laser or LED sources. Reprinted with permission from Axelrod (2001) CC-BY 4.0.

The EPI-to-TIR transition can also be easily seen in a sample of suspended fluorescence microbeads, as the out-of-focus background (in EPI) decreases to nearly zero and only the surface-proximal beads, all sharply focused, remain in view.

5.5.3 Azimuthal incidence angle control

The illumination of a sample by either TIR or highly oblique subcritical light often induces artifacts due to scattering and shadowing by refractive index variations in the sample. These artifacts are particularly noticeable with coherent laser illumination, which produces interference fringes arising from optical nonuniformities.

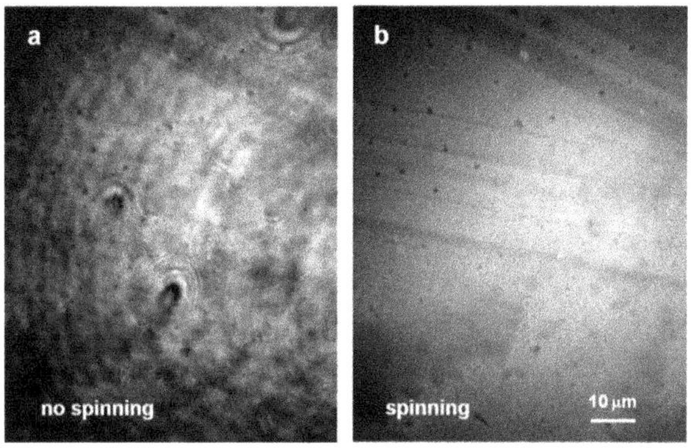

Figure 5.11. Effect of rapid azimuthal spinning on the laser-excited TIRF image of a diI-coated coverslip in water. Reprinted from Mattheyses et al (2006), copyright (2006) John Wiley & Sons.

One way to suppress their effect is to rapidly spin the azimuthal angle of illumination around the BFP. The artifacts do not disappear at any instant, but they are averaged during the camera exposure time, giving a better view of what is actually on the sample. Figure 5.11 shows the effect.

Azimuthal spinning is easily accomplished with double GMs by programming one mirror to oscillate its angle using time t and angular frequency ω according to the function cos ωt, and its orthogonal partner to oscillate according to sin ωt.

5.5.4 Switching excitation colors

Frequently, two (or more) fluorophores on a sample need to be followed simultaneously over time. 'Simultaneity' can be approximated by rapidly switching the excitation wavelength to excite one fluorophore or another, in synchrony with the camera exposures. Switching emission filter color optics is also possible, but somewhat slower and vibration-inducing, so it is better to use a single filter cube that has a complex band-pass spectrum which is appropriate for multiple fluorophores.

Implementing TIRF with multiple colors presents difficulties related to chromatic dispersion, but they are surmountable. In figure 5.9, the light incident upon L2 is either blue or green, depending on which shutters (SH1, SH2, or SH3) are open. The two color paths are combined at a dichroic mirror (DM1). However, the correct positioning of L2 and L3 was based on green light. The remaining lens in the system is L1, positioned exclusively in the path of the blue laser beam. This accounts for the fact that the L2 and L3 focal lengths are a little bit shorter for blue light than for green light. The green light that enters L2 is collimated. L1 focuses the blue light to a point a few centimeters in front of L2 in order to place its L2-refocused spot farther downbeam than would otherwise occur. The longitudinal position of L2 provides a handle to ensure that the blue beam's spot on the ceiling (after passing through the objective) is minimized in size.

5.5.5 Excitation polarization control

The ability to switch excitation polarization rapidly and repeatedly allows TIRF to measure molecular orientations and membrane folding, which will be discussed in the next chapter as applied to cellular secretion processes.

We want to produce either p-pol or s-pol on the sample, which is controlled by whether SH1 or SH2 are open. The first element encountered after the laser is a linear polarizer (LP). This is not necessary, since the laser output is already linearly polarized (vertical). However, by rotating the LP, the intensity at the angle of that linear polarization can be manually adjusted. The next element is a quarter-wave plate, which converts linear polarization into elliptical polarization and, at the optimal rotation angle, circular polarization. Circular polarization contains equal intensities of vertical (V) and horizontal (H) polarization.

The next element is the first PC. This sends the vertically polarized (V-pol) component off at 90° to the original direction. The horizontally polarized (H-pol) component continues straight on through. The next PC, after the shutters and mirrors, then recombines the H and V paths, now alternately illuminated according to the opening of the shutters. The H-pol light again goes straight through the second PC, and the V-pol light is deflected by 90° to rejoin it.

5.6 Alignment

In building a new custom TIRF system (because it has more open accessibility for modifications and new innovations than a prepackaged commercial system), or when realigning a grossly misaligned system, it is often best to realign the whole system starting from the 'beginning.' Perform the following steps in order (all with reference to figure 5.9)

(a) Position the GM duo set so that the last one in the path toward the microscope is at the center of its 'rear port' illumination path, which is a large circular hole in the microscope tower. Try to be accurate: you will need to use a ruler for this job.

(b) Using the riser dual mirror units that are positioned right in front of each of the lasers, set the beam to come out horizontally and at a height above the table equal to the height of the top mirror of the GM (use a ruler). This height should always apply to both colors, everywhere in the whole system.

(c) With no lenses in the path, adjust the mirrors (M) such that the H-pol green beam passes through the middles of the PCs and then along the table toward the middle of the top mirror in the GM, following a path that is parallel to the alignment of the screw holes in the optical table.

(d) With the objective removed, check that the beam comes through the middle of the hole below the objective screw mount. (Use a business card for this.) If it does not, then use the computer program controlling the GM to move the beam so that is the case. Now check whether the beam is going straight up. (Preferably, pre-mark this place on the ceiling). If the beam does not go straight up to the ceiling, then the overall GM position (not its mirror angle) is not exactly in the right spot. Adjust that, and go back to step (b).

By performing this GM adjustment, you will get a feeling for whether adjusting the overall GM position in any direction makes things better or worse. Continue repeating the steps (b)–(d) by correcting the GM in the direction that makes things better.

(e) Place pinhole diaphragms along the green path, marking where the raw H-pol beam passes.

(f) Now align the V-pol path to the H-pol path, judging by whether the V-pol goes through the center of two successive pinholes in the already-established H-pol path. The co-alignment is done by adjusting the mirror in the V-pol path between PC1 and PC2, and also the angle of PC1 using the thumbscrew on its mount. (Changing the angle of PC1 mainly affects V-pol (which is reflected) and not H-pol (which is transmitted)).

(g) Align the raw blue beam (no lenses yet) to be collinear with the green beam by adjusting the angle of the mirror just before L1 (which is not yet installed) and the angle of DM1.

(h) Now it is time to add the lenses. Install L2 so that the expanding beam it produces is still centered around the raw beam path. Install L3 so that the beam still emerges through the middle of the hole behind the objective screw thread mount. Install L1 so that the blue beam is still coincident with the green beam path after DM1.

(i) Make a sample dish. Put a droplet of diI in ethanol solution into a dish with a glass coverslip bottom and then immediately rinse it off with water. Now place a fluorescein solution in the dish to a depth of several millimeters.

(j) Screw the high-aperture objective (at least 1.45) into the microscope and place the sample dish on the stage. Now adjust the lateral position of L3 so that the beam comes through the objective and heads straight up.

(k) Focus the microscope on the orange diI fluorescence from the glass coverslip. If there is too much green fluorescein fluorescence to see the orange, dilute the fluorescein solution. The orange fluorescence should be centered in the field of view and centered in the squarish image of the GM. Refine L3's position to make the fluorescence look centered. Refine L2's axial position to make the beam spot on the ceiling as small as possible.

(l) Switch the shutters to the blue lasers and ensure that the fluorescence is centered. If it is not, tweak the angle DM1 slightly to make it so. Do not fiddle with L2 or L3 for this: they were already set with the green beam. Then, ensure that the blue spot on the ceiling is minimized in size. Adjust the axial position of L1 to make this happen.

(m) Now everything should be all set to adjust the computer-controlled GM toward TIRF. This happens when the fluorescence turns from mostly green (and out of focus, coming as it does from the solution) with a little surface orange, to almost pure orange, with evident sharp scratches and defects in the glass coverslip's surface.

5.7 Rapid chopping between TIR and epi-illumination

Regardless of the method chosen to produce TIR through a microscope, it is often desirable to rapidly switch between illumination of the surface (by TIR) and deeper illumination of the bulk (by standard epi-fluorescence, EPI). For example, a transient process may involve simultaneous but somewhat different fluorescence changes in both the submembrane and the cytoplasm, and both must be followed on the same cell in response to some stimulus. This can clearly be accomplished with an objective-based system that uses a GM set, as above. It is also possible to set up two different paths, one for TIR and one for EPI, in a prism-based system without a GM. Figure 5.12 shows a method for achieving rapid chopping using computer-driven AOMs.

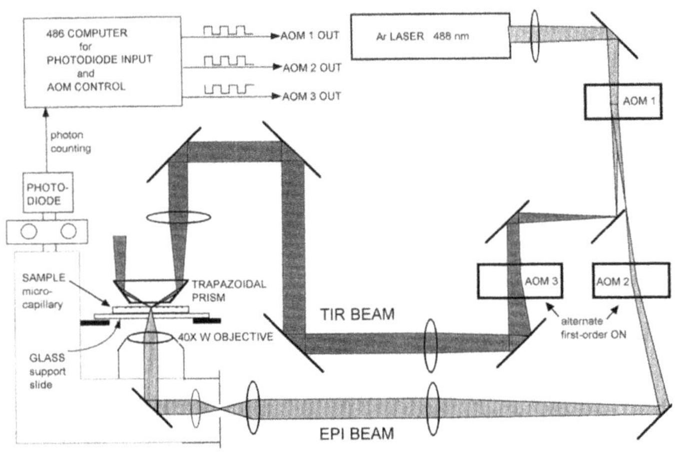

Figure 5.12. Optoelectronic system used for rapid chopping between TIR and EPI in a prism-based system, without the need for a GM set. The zeroth-order diffraction direction of acousto-optic modulator 1 (AOM1) is the TIR path; the first-order direction is the EPI path. AOM2 and AOM3 serve only to increase the on/off contrast of the EPI and TIR beams, respectively, so that they alternate cleanly. The computer that controls the square-wave pulses sent to the AOMs also acquires data from the single-channel photon counter so that the time courses of the two illumination modes can be separated in software. Reprinted from Omann and Axelrod (1996), copyright (1996), with permission from Elsevier.

5.8 Supercritical-angle fluorescence (SAF) emission setup

There are numerous ways to selectively block subcritical emitted light at a BFP and transmit only supercritical emitted light (to achieve further surface selectivity, as discussed in 4.11–4.13). Since the actual BFP is usually inaccessible (deep within the objective), the general approach is to use an EBFP. An EBFP resides 'upbeam' from the removable eyepiece; in fact, looking down through the eyepiece port gives a view of the EBFP. Figure 5.13 shows an actual photograph of that view, taken with a smartphone camera, using a fluorescent sample (discussed in section 5.5.2 above) that produces different emission colors for super vs subcritical light.

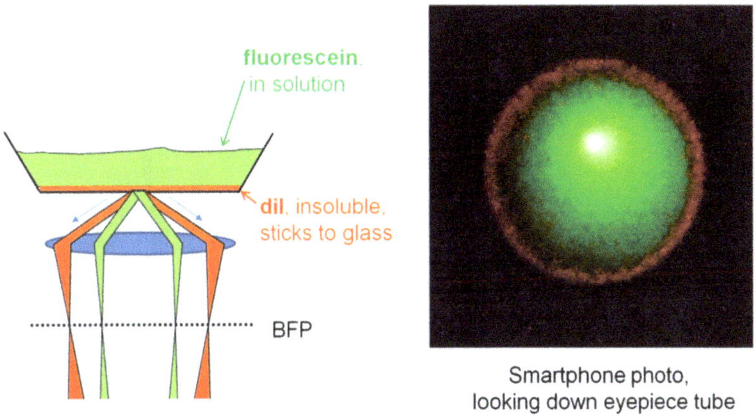

Figure 5.13. View of EBFP (with eyepiece removed), showing that orange emission from surface-adsorbed diI mainly goes into the high-radius supercritical zone, whereas solution-resident green fluorescence exclusively goes into the subcritical zone.

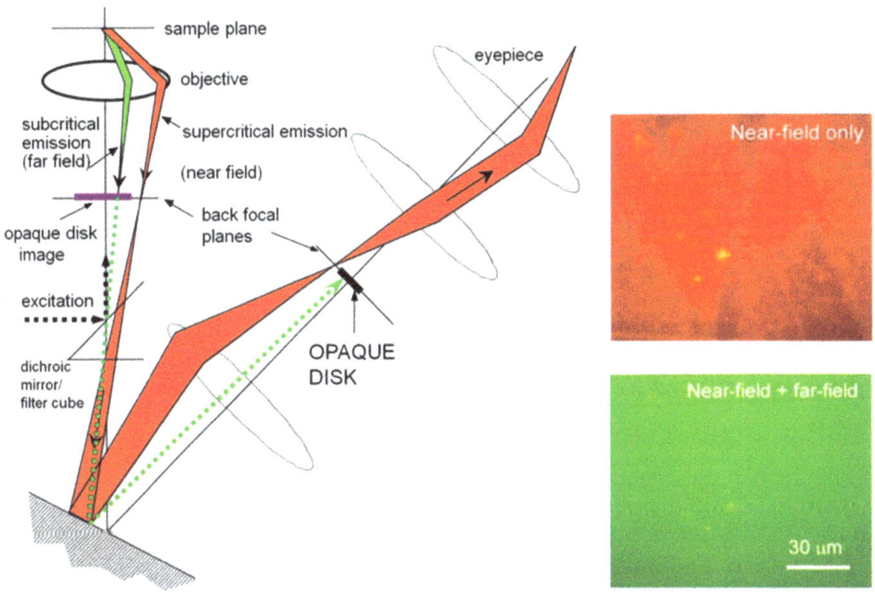

Figure 5.14. Blocking subcritical light (shown in green) at an EBFP near the eyepiece (or camera). Reprinted from Axelrod (2007), copyright (2007) with permission of Springer. The right-hand panels show the recorded images with and without subcritical blocking.

Figure 5.14 shows a schematic ray diagram for viewing only SAF, along with actual microscope views of the above fluorescein/diI sample.

Another possibility is to use a commercial beam splitter or image splitter, followed by lenses to reform the BFP in each of the two separated paths. Each path would then incorporate an opaque obstruction, one path blocking the

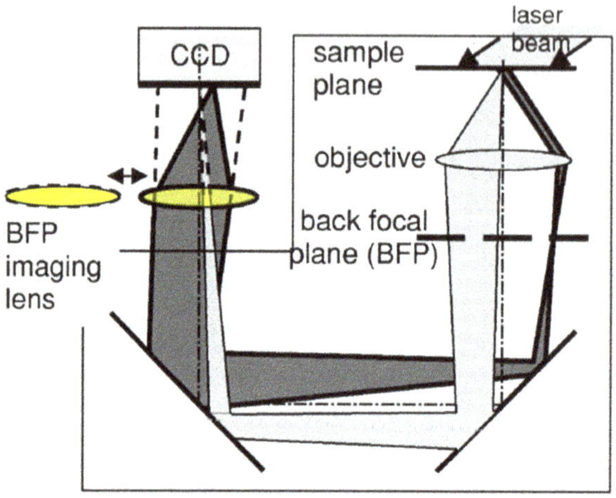

Figure 5.15. Optical modification used to record the intensity at the BFP directly. The path of typical parallel rays emanating from a range of sample points, but all at a particular angle, is shown with dark gray shading. The path of typical rays with a range of angles, but all emanating from a single point of the sample, is shown with light gray shading. Ray paths for the extra BFP lens (shown in yellow), when inserted, are shown by solid lines. The paths when the lens is removed are shown by dashed lines. This schematic of an inverted microscope's optical system is highly simplified; none of the fixed intermediate lenses and their consequent intermediate focal planes after the objective are depicted. The placement of these intermediate elements varies by manufacturer, e.g. finite versus infinite tube length, and inverted versus upright configuration. Reprinted with permission from Mattheyses and Axelrod (2005) CC-BY 4.0.

subcritical, and the other blocking the supercritical light. Two separate detection systems would observe the throughputs.

5.9 Imaging the back focal plane directly

The BFP is a special place in a microscope, where the angles of excitation and emission at the sample (crucial to TIRF and SAF microscopies) appear transformed into positions. As such, the BFP contains information about the z-positions and orientations of fluorophores, possible dielectric films, and surface plasmon effects from metal films. The intensity pattern of the BFP image can be recorded using the microscope's high-quality CCD camera system by simply inserting an extra lens a few centimeters before the CCD, as shown in figure 5.15.

With the BFP imaging lens in place, the microscope's BFP is imaged at the face of the CCD array. With the BFP imaging lens removed from the path, the microscope's SP is focused at the face of the CCD array. The extra BFP lens serves the same function as replacing the eyepiece with a 'Ramsden eyepiece,' which is often used to superimpose and align the images of phase contrast rings at the BFP.

Figure 5.16 shows actual images taken with such a system on a diI-coated surface. It contrasts the BFP for a diI/bare glass surface with the BFP for a diI/metal film/glass surface.

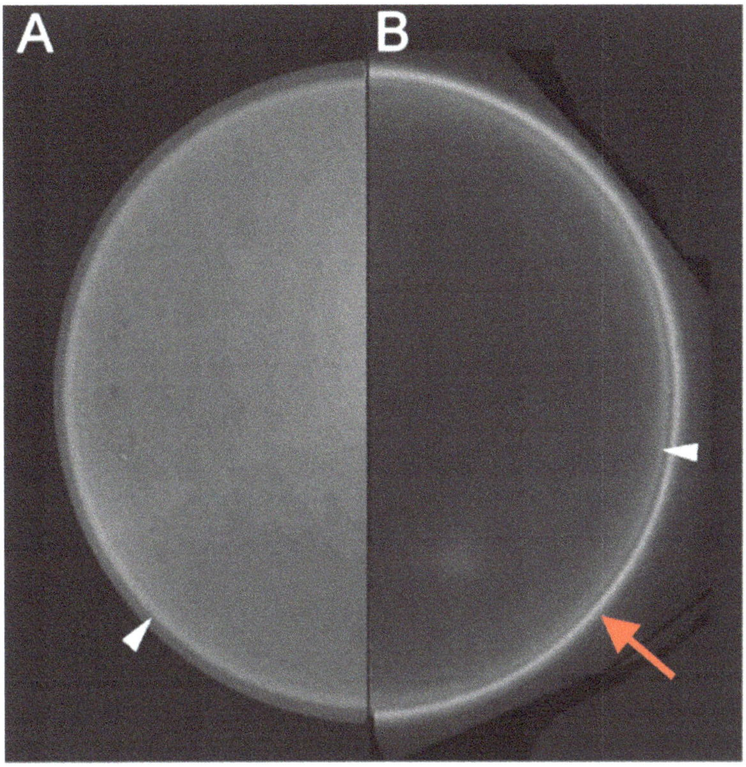

Figure 5.16. Experimental composite of images of the BFPs of two different samples. (A) diI on bare glass. The white arrowhead indicates the intensity peak at the emitted light's critical angle, as discussed in chapter 4. (B) diI on a 20 nm aluminum coating on a glass coverslip. The white arrowhead indicates the intensity peak at the critical angle at the same radius as on the bare glass, and the full arrow (in red) indicates the peak at the surface plasmon angle, at a larger radius and thus a larger emission angle. Objective NA = 1.4. Reprinted with permission from Mattheyses and Axelrod (2005) CC-BY 4.0.

5.10 Measurement of evanescent field depth

Experimentally measuring the actual evanescent field depth d of a sample is not quite as straightforward as it seems it should be, because many methods involve perturbing the field (say, with a microprobe), which may then change its profile. In addition, scattering and a spread of supercritical incidence angles complicate the measurement.

One method is to directly measure the polar angle of incidence θ_3 of the excitation beam in the coverslip, assume that it is well collimated, and then deduce d from equation (2.12). The problem with this approach is that the incident beam path is quite inaccessible, especially with objective-based TIRF. However, the incident beam can be 'coupled' out into a propagating beam that falls upon some distant surface (such as a lab wall), using a prism of known refractive index and shape (see figure 5.17).

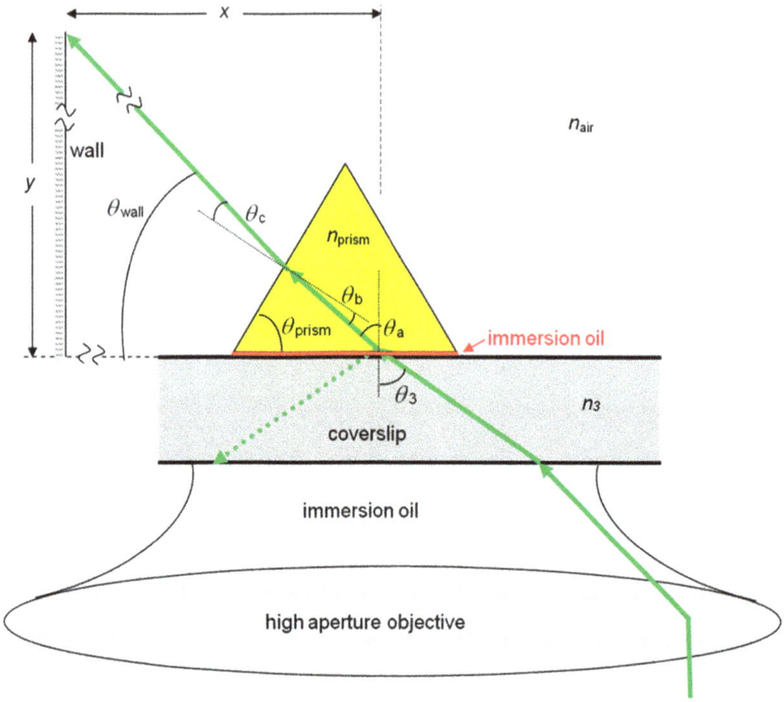

Figure 5.17. Scheme for indirectly measuring the actual incidence angle θ_3 by coupling the incidence beam out with a prism (shown in yellow). The beam path without the prism is shown by a dashed line.

$\theta_{wall} = \tan^{-1}(y/x)$ can be obtained from the measured distances x and y. The incidence angle θ_3 and d can then be computed using Snell's Law and geometry.

Another way to measure d involves placing an index-matched spherical bead (with a radius of ~10 μm) on the TIR surface. The bead should be pre-labeled with a fluorescent dye on its surface (e.g. diI). As can be seen in figure 5.18, TIR illumination excites only the near-contact region, and the fluorescence intensity becomes dimmer with radial distance r in a microscope image. The dependence of the intensity upon r can be used to calculate the evanescent field depth d. This calculation assumes that the evanescent field is indeed exponentially decaying and that the index matching is close to perfect.

A third way to measure d is the most indirect of all, but it does not perturb the evanescent field. Under TIR illumination, a dilute solution of a fluorescent material (molecules or very small beads) appears as twinkling and jittering specks of light, as the solute diffuses in and out of the evanescent field depth by Brownian motion. The temporal behavior of the intensity fluctuations in any small lateral region can be autocorrelated to determine the average fluctuation rate. This rate is proportional to $1/d^2$. The proportionality factors can be derived using the methods given in section 6.8.2, which discusses TIR/fluorescence correlation spectroscopy (FCS).

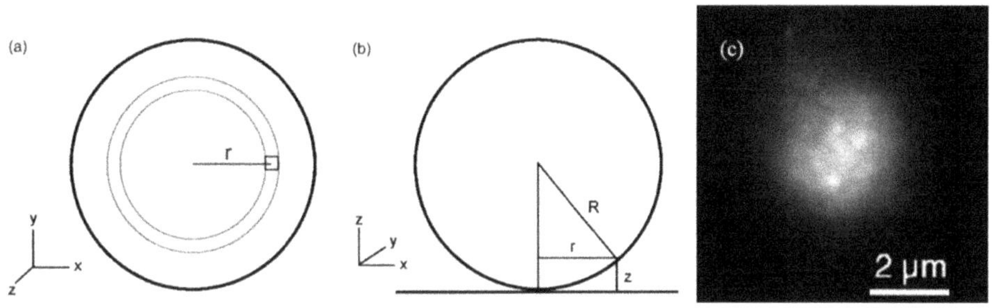

Figure 5.18. Scheme for measuring *d* using a surface-labeled spherical bead in the evanescent field. (a) View of the bead in the *x*–*y* plane as observed though the microscope. The ring outlined is one radial pixel bin wide, corresponding to a radius *r*. One square pixel in this ring is shown. (b) View of the cross section of a bead resting on a coverslip in the *x*–*z* plane, showing the conversion of a projected *r* radius into a *z* distance from the coverslip. (c) An actual image made using this procedure. The visible 'splotch' of fluorescence is only that portion of the bead in close contact with the coverslip; the full bead is much larger. The irregular pattern is due to nonuniformities in the surface labeling, which thereby necessitate azimuthal averaging at each radius. Reprinted with permission from Mattheyses and Axelrod (2006) CC-BY 4.0.

5.11 TIRF–structured illumination microscopy (TIRF–SIM)

In all of the previous sections, it was implicitly assumed that the evanescent intensity has no important *x*–*y* dependence; the illumination was uniform across the usable field of view. (Recall, however, that the area of illumination is somewhat adjustable by changing the width of the incoming light beam.) However, introducing a finer-scale *x*–*y* pattern—a 'structure'—has many uses, ranging from measuring lateral diffusion to enhancing the resolution. Three different patterns will be discussed here: (a) single-spot TIR; (b) multispot TIR; and (c) periodic-pattern TIR. Some of these approaches essentially combine TIRF with the standard EPI-illumination techniques of confocal microscopy and periodic structured illumination microscopy (SIM). The pattern-producing illumination schemes all involve simultaneous illumination from multiple directions.

To understand how an objective focuses light to a spot on the SP, and how the objective views fluorescence from that spot, we must first review some essential theory in Fourier optics.

5.11.1 Single-spot TIR with converging illumination

The setup shown in figure 5.8, with its azimuthally isotropic illumination, produces a broad area of TIR illumination with rays that are all at the same polar incidence angle, which ensures a single exponential evanescent field profile in the *z* direction. This results from a very thin ring of illumination focused at the BFP.

These optics can be modified to achieve the opposite extreme (figure 5.19): a small area of illumination, but with a range of polar incidence angles and evanescent field decay depths.

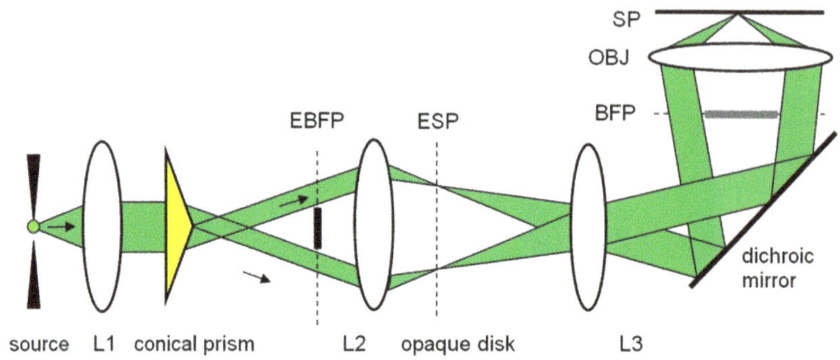

Figure 5.19. Production of spot TIR illumination with converging, incoherent, and unpolarized light. Reprinted with permission from Mattheyses and Axelrod (2006) CC-BY 4.0.

The optical elements in figure 5.19 are essentially similar to those of figure 5.8, but with their positions and focal lengths modified so that the beam is collimated (rather than focused) as it passes the BFP. This ensures that the beam is instead focused at the sample. Nonetheless, any rays must be blocked that might pass through the BFP at small enough radii to be subcritical at the sample. This cannot be done at the actual BFP, because that position is inside the objective, and also we do not want fluorescence to encounter that blockage. Instead, the opaque disk is emplaced at a position where lenses L2 and L3 create an EBFP. A 'sharp shadow' of that disk will then form at the actual BFP, excluding all subcritical rays. The reason that a ring of illumination, rather than a single beam, is best for the above scheme is to ensure that the final spot is azimuthally symmetric.

The questions are: how small is the TIR spot on the sample? What is its profile? The electric field at the SP (the focal plane of the objective) is exactly the Fourier transform of the field at the BFP (Goodman, 1968, equation 5–19). We can use the Fourier transform solution already derived (equation 4.46) for a lens uniformly illuminated up to some radius R. In that case, we were not looking at incident light heading towards the sample, but rather at the fluorescence focused to a spot by the microscope tube lens. Equation (4.46), rewritten here, is:

$$E'(r') = E_0 \frac{R}{r'} J_1\left(\frac{kR}{f} r'\right). \tag{5.5}$$

In the present case, f refers to the focal length of the objective, not the tube lens. The phase factor α that appears in equation (4.46) is exactly 1.0 here, which is appropriate to focus the electric field pattern at the BFP onto the SP. Furthermore, the definite r-integral in equation (4.44) leading up to equation (4.46) should range—not from zero—but from radius R_1 (corresponding to the minimum aperture in the supercritical range that is not blocked by the opaque disk shadow) up to radius R_2 (corresponding to the maximum aperture of the objective). That integral from R_1 to R_2 is just the difference between the integrals covering $(0,R_2)$ and $(0,R_1)$:

$$E'_{\text{ring}}(r') = E_0 \frac{1}{r'} \left[R_2 J_1\left(\frac{kR_2}{f} r'\right) - R_1 J_1\left(\frac{kR_1}{f} r'\right) \right]. \tag{5.6}$$

For an objective that follows the Abbe sine condition, $R_i/f = \sin\theta_i$, where θ_i is the angle of convergence toward the focal spot. Here, k refers to the wavenumber in the immersion medium (not the vacuum) between the objective and the coverslip (presumably index matched at n_3 to the coverslip), so $k = 2\pi\, n_3/\lambda_o = k_o n_3$.

The intensity pattern $I'_{\text{ring}}(r')$ at the SP focus arising from supercritical ring illumination of the BFP is the square of the amplitude of $E'_{\text{ring}}(r')$:

$$I'_{\text{ring}}(r') = E_0^2 \frac{1}{r'^2} \left[R_2 J_1\left(\frac{2\pi n_3 \sin\theta_2}{\lambda_0} r'\right) - R_1 J_1\left(\frac{2\pi n_3 \sin\theta_1}{\lambda_0} r'\right) \right]^2 \tag{5.7}$$

The factors $n_3 \sin\theta_i$ are just the effective numerical apertures A_i (called NA in the previous sections) of the objective at the inside and outside radii of the supercritical light ring at the BFP. Equation (5.7) can be rewritten as:

$$I'_{\text{ring}}(r') = E_0^2 \frac{R_2^2}{r'^2} [J_1(k_0 A_2 r') - \beta J_1(\beta k_0 A_2 r')]^2, \tag{5.8}$$

where β is defined as the ratio of the numerical apertures, i.e. A_1/A_2.

Figure 5.20 shows the normalized shape of $I'_{\text{ring}}(r')$, comparing supercritical ring illumination (black) in the A range (1.40–1.49) to full-aperture illumination (red) in the A range (0–1.49).

Restricting the illumination aperture to the TIR-producing supercritical ring actually makes the radius to the first minimum shorter than it is for (mostly non-TIR-producing) full-aperture illumination; i.e. the central maximum becomes sharper. However, rings become brighter at higher radii, which decrease overall sharpness. An image-plane pinhole (analogous to that normally used in confocal microscopy) could be installed to block the fluorescence excited by those rings.

The optical setup in figure 5.19, coupled with an image-plane diaphragm for the emitted light, is the TIR analog of the confocal microscope: a scanning spot

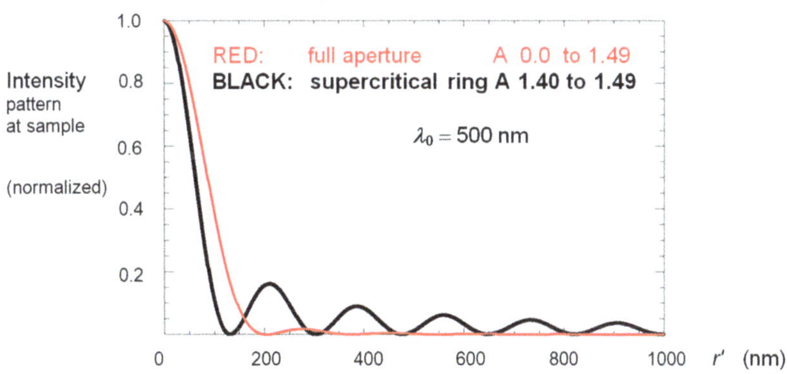

Figure 5.20. Intensity pattern at the SP for supercritical ring illumination vs full-aperture illumination.

(or stage), a focused spot, and an image-plane pinhole. Standard confocal microscopy's main purpose is to reduce out-of-focus axial light. The purpose of the analogous TIR-confocal system is to improve lateral resolution (since out-of-focus light is already minimized by TIR).

The image-plane pattern for supercritical ring illumination (equation 5.8) is not actually the evanescent field pattern; rather, it is the intensity profile of the incident hollow cone at its vertex. The derivation completely ignores the presence of an interface at that position. The evanescent field should be very similar to the incident light pattern, but three effects will ensure that equation (5.8) is only an approximation. First, the evanescent field is always shifted slightly from the incident field by the Goos–Hänchen effect (see section 2.4). Second, the intensity of the evanescent field is a decreasing function of incidence angle (see figure 2.2), an effect which requires some relative weighting among the range of polar incidence angles in this TIR configuration. The weighting here is implicitly assumed to be unity through the polar angle range. Third, the derivation ignores polarization.

5.11.2 Array of TIR spots

The single-spot TIR scheme depicted in figure 5.19 can be developed further into a multispot TIR array, which can then be scanned across the sample in confocal-microscope fashion to fill out a whole field of view. The multispot feature allows the scan to cover a smaller area (and take less time). The result is a full-field view with an x–y resolution which exceeds that of a standard wide-field TIR because of the narrow central peaks in the illumination produced by each spot (see figure 5.20).

The principle behind the scheme is simple: create multiple beams with a microlens array, consisting of perhaps thousands of tiny lenses. Almost all of the lenses are off-axis and thereby create separate off-axis beams converging to their own separate foci. A lens (L1 in figure 5.21) collimates these beams. The collimated beams all travel from different directions toward the optical axis, but they all cross at the focal plane of L1. The succeeding lenses L3 and L4 assure that the focal plane of L1 is complimentary ('equivalent') to the objective's BFP, an 'EBFP.' At the EBFP, an opaque disk is installed, thereby casting a sharp shadow at the actual BFP. The radius of the disk at the EBFP is chosen to block the part of the light from each collimated beam that was heading toward the subcritical zone of the BFP; only the part of each collimated beam heading toward the supercritical zone at the BFP is allowed to pass. The objective then refocuses the supercritical light of each collimated beam to its own spot, an image of the ESP planes between lenses L3 and L4 (and of the spot array just after the microlens array). Replacing the L3–L4 combination with a single lens serves the same purpose, but two lenses give the system more flexibility to access the ESP and to adjust the final spacing of the spots on the sample.

Figure 5.21 shows only the illumination portion of the system. An array of matching pinholes needs to be installed in the fluorescence emission beam in order to block the sidebands shown in figure 5.20. Scanning is accomplished by a GM or an

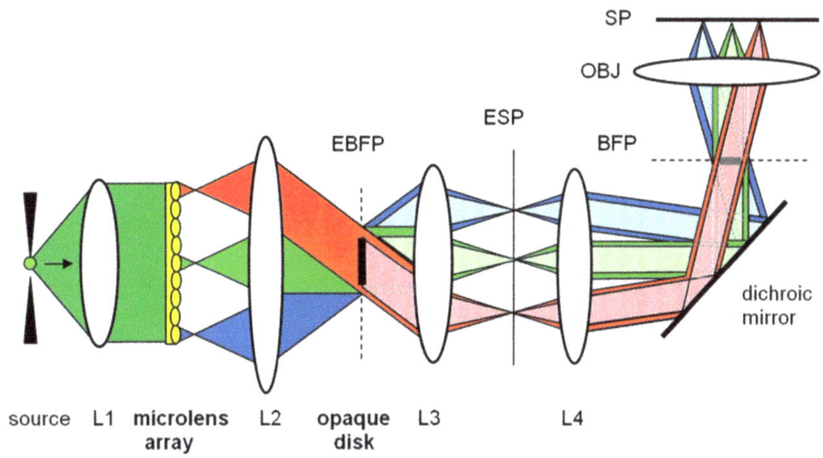

Figure 5.21. Scheme for producing multispot TIR in a confocal-like system. Here, only the excitation path and the aspects that are unique to TIR production are shown. The GM scanners in the excitation path and the corresponding pinhole array in the emission path are not shown; they are similar to those in a standard confocal microscope. In this figure, the colors are used only for pictorial clarity: all the beams are indeed the same actual color. Only three beams (out of thousands in a 2D array) are shown for simplicity (depicted in red, green, and blue). The regions of each where light has been blocked by the opaque disk are shown as a much lighter version of their parent color; in reality, those regions are devoid of light. The image or 'shadow' of the actual opaque disk (black) is shown as a grey line at the BFP.

electro-optical deflector (not shown) installed in the excitation path. The action of the GM allows the beams to 'pivot' their angles at the BFP so that the multispot array moves, but the sub- vs supercritical positions at the BFP do not move. Just as in a single-spot confocal microscope, the pinholes in the emission path do not need to be moved or scanned at all, because the sample fluorescence spots are viewed in the reflection of the same GM that scans the excitation spots, at a plane created to be an ESP. For an example of the details, see Guo *et al* (2018).

A disadvantage of the above multispot TIR system is that a lot (indeed most) of the source light is 'thrown away' by being blocked at the opaque disk. However, a sufficiently bright LED source is acceptable. The light hitting the microlens array must be collimated so that the spots it produces are not laterally smeared out.

5.11.3 Periodic sine-wave pattern

Using coherent light intersecting from multiple directions, an interference pattern forms in the evanescent field, and this can have important applications. One application is for image resolution enhancement (discussed in this section); another is in fluorescence recovery after photobleaching studies of surface diffusion (discussed later in section 6.8). The optical setup features two collimated laser beams split from the same laser source (figure 5.22).

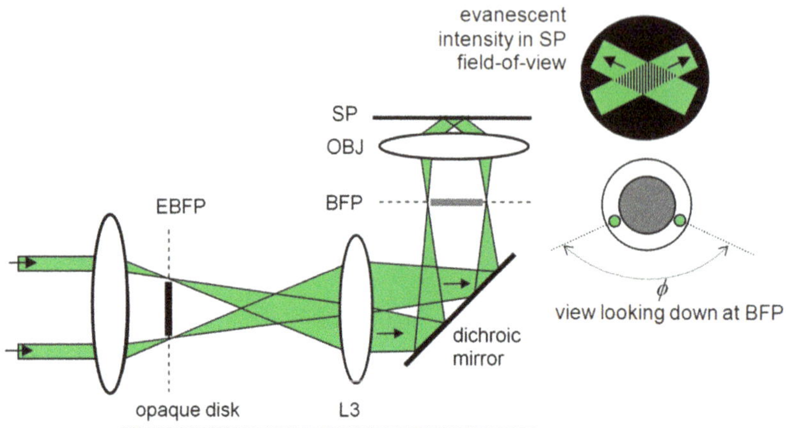

Figure 5.22. Optical setup that uses two mutually coherent laser beams to create a stripe pattern.

The evanescent electric field in medium 1 is sinusoidally periodic in the x–y plane (the plane of the interface supporting the TIR), with a slight phase shift relative to the incident propagating plane wave in medium 3 that is incidence-angle dependent. The phase of the evanescent electric field has no z dependence, so the wavefronts are perpendicular to the interface and the field can be represented simply as a 2D wave in x–y. Because of the close spacing of wavefronts in the evanescent field, illumination by a pair of intersecting and mutually coherent TIR laser beams can produce a very closely spaced striped interference fringe pattern in the evanescent intensity (provided that the beams have the same evanescent polarization). If the relative azimuthal angle between the two TIR beams at the BFP is ϕ, then the node-to-node spacing s of the evanescent fringes is given by

$$s = \frac{\lambda_0}{2n_3 \sin \theta \sin (\phi/2)}, \tag{5.9}$$

where θ is the polar incidence angle and λ_0 is the vacuum wavelength. The spacing s is not dependent upon the refractive index n_1 of the medium (or cell).

Figure 5.23 shows an example of intersecting coherent TIR beams creating a stripe pattern by mutual interference. The setup used for this particular example is actually a prism-and-mirror system as depicted in figure 5.1D.

5.11.4 Periodic pattern for image enhancement

By illuminating a sample with a periodic stripe pattern, along with a nontrivial image analysis protocol, the effective lateral resolution of an image can be enhanced. This is true for both non-TIR and TIR-based illumination, but the enhancement is marginally better for TIR. This section discusses why this is so.

Qualitatively, why does periodic SIM enhance resolution at all? In the periodic SIM scheme, three images are taken with the phase of the grid shifted 120° from one

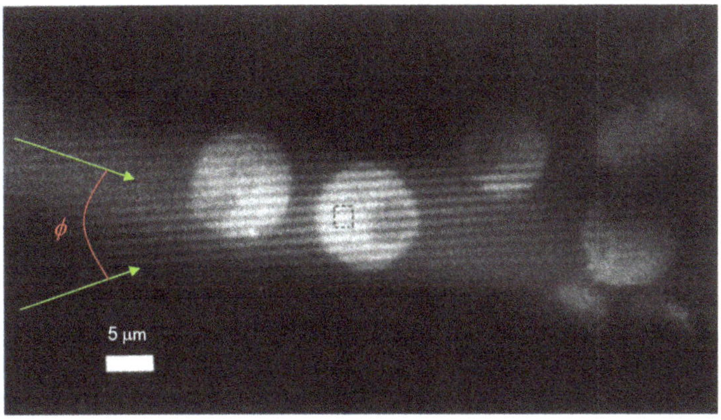

Figure 5.23. Flattened erythrocyte ghosts under TIR intersecting-beam illumination. The ghosts are labeled with adsorbed fluorescein insulin, with which they are in equilibrium in solution. The green arrows show the direction of travel of the wavefronts of the two intersecting evanescent fields. The relative azimuthal angle ϕ is rather small here. If it were at its maximum possible of 180°, the fringe spacing would be less than the resolution. While it would still exist at full contrast, it would be virtually invisible as seen through the microscope objective. Adapted from Fulbright and Axelrod (1993), copyright (1993) Plenum Publishing Corporation, with permission of Springer.

image to the next. Clearly, a tiny fluorescent object that might be in a node in one image will not be in the other two, and a comparison of the three images clearly defines the location of that tiny object to an accuracy much better than the microscope's resolution, at least if the periodicity of the illumination is finer than the standard resolution of the microscope.

Quantitatively, the standard resolution is given by the Raleigh resolution limit, equation (4.50), which assumes that two infinitesimal objects can be resolved (just barely) if the central maximum of one of their Airy patterns falls on the 'first zero' of the other's Airy pattern. (As mentioned, this is an approximation that ignores near-field emission and polarization.) For an NA = 1.65 objective (close to the highest commercially available NA) and a vacuum wavelength λ_0, the Raleigh limit is then $s_{ral} = 0.61\lambda_0/\text{NA} = 0.37\lambda_0$. This corresponds to a maximum wavenumber $k_{ral} = 2\pi/s_{ral} = 17.0/\lambda_0$.

Compare this to the maximum wavenumber achievable by a periodic pattern set up by opposing intersecting beams (ϕ = 180° in equation 5.9). If we limit the maximum polar angle to the critical angle (say, for a glass/water interface), so that the interference pattern is still EPI (non-TIR but with the opposing beams skimming the surface), then set $n_3 \sin\,\theta = 1.33$, equation (5.9) then yields a node-to-node periodic spacing of $s_{EPI} = 0.38\lambda_0$. This corresponds to a periodic-pattern wavenumber of $k_{EPI} = 2\pi/s_{EPI} = 16.5/\lambda_0$.

According to periodic SIM theory (Gustafsson *et al* 2008), the effective maximum wavenumber becomes the sum of these two wavenumbers, $k_{sum} = k_{ral} + k_{per} = 34.0/\lambda_0$. Therefore, the EPI (non-TIR) SIM technique (using the highest available NA objectives for each) approximately doubles the effective resolution. If, instead, the opposing beams are allowed to exceed the critical angle, and instead use the supercritical objective-based extreme of NA = $n_3 \sin\theta = 1.65$, then $s_{TIR} = 0.30\lambda_0$. This corresponds to $k_{TIR} = 2\pi/s_{TIR} = 20.9/\lambda_0$, and k_{sum} improves to $36.2/\lambda_0$, an increase of about 6% over the best resolution achievable by non-TIR SIM.

In any periodic SIM experiment (EPI or TIR), it is necessary to record at least three images, usually with relative 120° phase shifts in the periodic pattern for each pair of opposing beams. The three required relative phase shifts can be easily achieved by the insertion of a one-third-wave retardation plate of the appropriate index and thickness into one of the two beams that were split from the original laser beam, anywhere upstream in the optical path. However, to achieve improved resolution uniformly in all lateral directions, at least three azimuthal orientations of opposing beams (say at 0°, 60°, and 120°) should also be employed, making a total of nine images of each sample scene.

5.11.5 Spot TIR with collimated light

Collimated incident light has the advantage of utilizing just a single incidence angle and thereby a single evanescent exponential decay depth. In general, collimated incident light produces a wide area of illumination, i.e. the same area as an oblique section through the beam. However, it is possible to retain the 'clean' single incidence angle while producing a very small TIR spot.

The idea of the intersecting coherent evanescent fields used to generate the sine-wave periodic pattern discussed above can be carried farther by introducing more directions. For example, four beams arranged at a spacing of 90° around the BFP produce a checkerboard pattern. Eight beams (see figure 5.24(A)) produce a pattern which is a crystalline-like array whose center is the brightest, because this is where all the incoming waves are exactly in phase. The number of incoming beams can be increased to 'infinity,' which is basically a coherent circle of waves converging to the center (figure 5.24(B)). Note, however, that complete interference will only occur if the light arriving from all directions has the same polarization. This can be achieved only if the evanescent field from each direction is z-polarized at the TIR surface, which can only occur if the ring of light at the BFP is *radially* polarized (figure 5.24(C)). Commercially available devices exist which can produce a ring of radially polarized light from a collimated linearly polarized laser beam.

Mathematically, an infinite number of 2D plane waves converging into circular wave can be written as a smooth integral over all azimuthal angles ϕ of plane waves of equal amplitude and equal phase (zero) at the center:

$$I(r) = \left[E\int e^{i\mathbf{k}\cdot\mathbf{r}}d\phi\right]^2 = \left[E\int e^{ikr\cos\phi}d\phi\right]^2 = I(0)[J_0(kr)]^2, \qquad (5.10)$$

where J_0 is a zeroth-order Bessel function. The intensity is not an infinitesimal point, because although the integration is done over all directions ϕ, it is limited to a very particular amplitude of k corresponding to the fixed wavelength of the circular

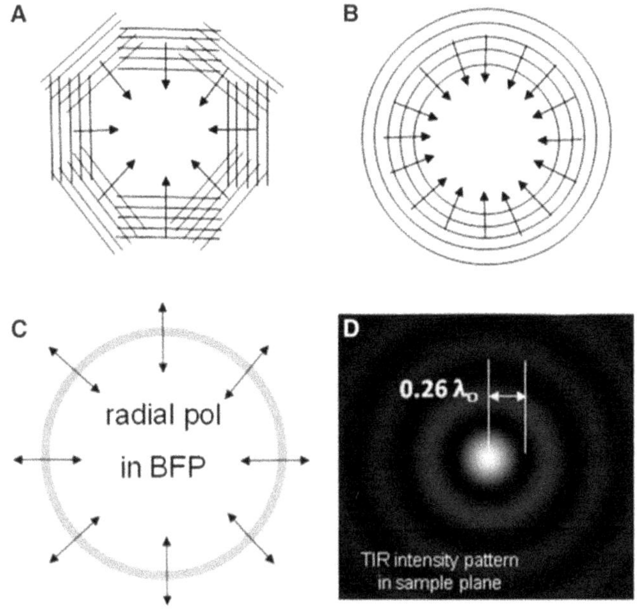

Figure 5.24. (A,B) Supercritical coherent illumination from increasing numbers of directions. If (C) the pattern on the BFP is a complete radially polarized ring, a very small bright central maximum of evanescent illumination (D) is created on the sample. Reprinted from Axelrod (2013), copyright (2013), with permission from Elsevier.

waves. (If all amplitudes of k were indeed present, then the integration would produce a 2D Dirac delta function.)

The single-spot illumination obtained using a circular TIR illumination pattern is similar, but not identical, to an Airy disk shape of the point spread function (PSF), as discussed earlier. In this case, the pattern is a zeroth-order Bessel function. The form of the Airy disk (equation 4.47) is not the same, but the two patterns can be related by the identity:

$$\frac{J_1(u)}{u} = \frac{J_2(u) - J_0(u)}{2} \tag{5.11}$$

The spot, while not infinitesimal, is very small by conventional microscopy standards. The $I(0)/e^2$ point resides at the Bessel function argument kr~1.75. Since the wavenumber corresponding to the evanescent wavefront spacing is

$$k = \frac{2\pi n_3 \sin\theta}{\lambda_0}, \tag{5.12}$$

the radius out to the $1/e^2$ point is ~0.18 λ_0. The radius out to the first minimum is ~ 0.26 λ_0. In comparison, the standard calculation for microscope resolution with an NA = 1.49 objective yields ~0.41λ_0. Clearly, the evanescent spot radius produced on the sample by a radially polarized ring at the BFP is about 50% smaller than the best

Figure 5.25. Lateral view of microvolume produced by interference of a converging circle of evanescent light, z-polarized all around, through an objective with NA = 1.49, in an aqueous solution of fluorescent dye, $n_3 = 1.515$. The evanescent depth d is calculated to be < 100 nm.

resolution of a microscope. By raster scanning (presumably, scanning the sample so as not to disrupt the perfect alignment of the incoming phases), lateral super-resolution can be achieved over a whole sample field. As in confocal microscopy, an image-plane diaphragm can be used, not to exclude out-of-focus light as in confocal microscopy, but rather to block light excited by the secondary rings. In addition, the z-resolution is also extraordinarily good (<100 nm), since the spot is evanescent. This is an improvement over confocal *z*-resolution by a factor of about five. An additional feature of this system is that the illuminated volume is extremely small, which makes fluorescence fluctuations (as in FCS) relatively large as single fluorophores enter or leave the illuminated region.

The alignment has to be perfect (and the quality of the optical elements very high) to achieve such a small volume. Nonetheless, figure 5.25 shows the TIR spot that can be experimentally achieved with 'ordinary' effort and standard optical elements.

Further reading

Axelrod D 1999 Surface fluorescence microscopy by evanescent wave illumination *Light Microscopy in Biology - A Practical Approach* 2nd edn ed A J Lacey (Oxford: Oxford University Press) 390–413

Axelrod D 2001 Selective imaging of surface fluorescence with very high aperture microscope objectives *J. Biomed. Opt.* **6** 6–13

Axelrod D 2001 Total internal reflection fluorescence microscopy in cell biology *Traffic* **2** 764–74

Axelrod D 2003 Total internal reflection fluorescence microscopy in cell biology *Methods Enzymol.* **361** 1–33

Axelrod D 2007 Total internal reflection fluorescence microscopy *Optical Imaging and Microscopy: Techniques and Advanced Systems* 2nd edn ed P Török and F-J Kao (Springer Series in Optical Sciences) (Cham: Springer) pp 195–236

Axelrod D 2013 Evanescent excitation and emission in fluorescence microscopy (Invited review) *Biophys. J.* **104** 1401–9

Axelrod D 2016 Total internal reflection fluorescence microscopy *Encyclopedia of Cell Biology vol 2 Organizational Cell Biology* (Amsterdam: Elsevier) 62–9

Fulbright R M and Axelrod D 1993 Dynamics of nonspecific adsorption of insulin to erythrocyte membranes *J. Fluor.* **3** 1–16

Goodman J W 1968 *Introduction to Fourier Optics* (New York: McGraw-Hill)

Gortari N G, Bouchoule S, Cambril E, Cattoni A, Hauke L, Enderlein J, Rehfeldt F and Yacomotti A 2020 Metasurface-based total internal reflection microscopy *Biomed. Opt. Express* **11** 1967–76

Guo M *et al* 2018 Single-shot super-resolution total internal reflection fluorescence microscopy *Nat. Methods* **15** 425–8

Gustafsson M G L, Shao L, Carlton P M, Rachel Wang C J, Golubovskaya I N, Cande W Z, Agard D A and Sedat J W 2008 Three-dimensional resolution doubling in wide-field fluorescence microscopy by structured illumination *Biophys. J.* **94** 4957–70

Mattheyses A L and Axelrod D 2005 Fluorescence emission patterns near glass and metal-coated surfaces investigated with back focal plane imaging *J. Biomed. Opt.* **10** 054007

Mattheyses A, Shaw K D and Axelrod D 2006 Effective elimination of laser interference fringing in fluorescence microscopy by spinning azimuthal incidence angle *Microsc. Res. Tech.* **69** 642–7

Mattheyses A and Axelrod D 2006 Direct measurement of evanescent field profile and depth in TIRF microscopy *J. Biomed. Opt.* **11** 014006

Omann G M and Axelrod D 1996 Membrane proximal calcium transients in stimulated neutrophils seen by total internal reflection fluorescence *Biophys. J.* **71** 2885–91

Chapter 6

Applications of TIRF microscopy and its combination with other fluorescence techniques

Subsections 6.8.1, 6.8.2, 6.8.3, 6.8.6 and 6.11.4 *of this chapter are coauthored with Dr Nancy L Thompson, Dept. of Chemistry, University of North Carolina, Chapel Hill. Subsection* 6.8.5 *of this chapter is coauthored with Dr Geneva M Omann, Dept. of Biological Chemistry, University of Michigan, Ann Arbor*

Aside from the most straightforward application of total internal reflection fluorescence (TIRF) microscopy in cell and molecular biology, which is to qualitatively reduce out-of-focus background, numerous applications of TIRF microscopy have drawn upon unique total internal reflection (TIR) properties and the combination of TIRF microscopy with other techniques. In this chapter, we limit the discussion of these other techniques to their underlying optical system principles and image analysis protocols, rather than an exhaustive review of experimental results.

6.1 Refractive indices in cell cultures

Here we discuss a TIR microscopy method to measure the laterally variegated refractive index of cell cultures. It is not a fluorescence technique in itself, but the technique's optical setup can be completely compatible with that of objective-based TIRF, such that the same field of view of the same sample can be examined using the same objective; it produces a lateral resolution down to about 1 μ. The results then can be used to predict the amplitude and pattern of TIR scattering, which decreases the purity of surface selectivity in TIRF microscopy, as discussed in mathematical detail in section 3.5. The technique is non-perturbative, requires no labeling, uses only one illumination wavelength, and requires only the standard cell buffer medium without replacement.

doi:10.1088/978-0-7503-3351-1ch6

The fundamental idea is that locations on the sample that have different refractive indices have different TIR critical angles, which should be detectable by varying the incidence angle and monitoring the intensity of the reflection. Figures 3.6 and 3.7 (and associated text) display the theory of what might be expected.

The optical configuration is based upon a custom objective-based (numerical aperture (NA) 1.49) TIRF system but with the standard fluorescence dichroic mirror/barrier filter cube removed and replaced with a fully reflective mirror (and no barrier filter) that only extends across a small fraction (15%–30%) of the aperture on one side (see figure 6.1). This ensures that the incident beam deflects up into the objective (and focuses at the back focal plane (BFP)), but the reflected beam from the sample passes back down on the opposite side of the aperture unimpeded. This configuration prevents light scattered from the first deflection of the beam toward the sample from reaching the camera, but in our implementation, the incident beam approaches the sample from one azimuthal angle only.

Despite the fact that it is 'looking' straight into the reflected beam while situated at an equivalent sample plane, the CCD does record in-focus intensity variations that are generated at the sample. The beam is scanned through a range of incidence angles that includes the range of TIR critical angles likely to be found on the variegated sample. Because the beam does not utilize much of the objective aperture,

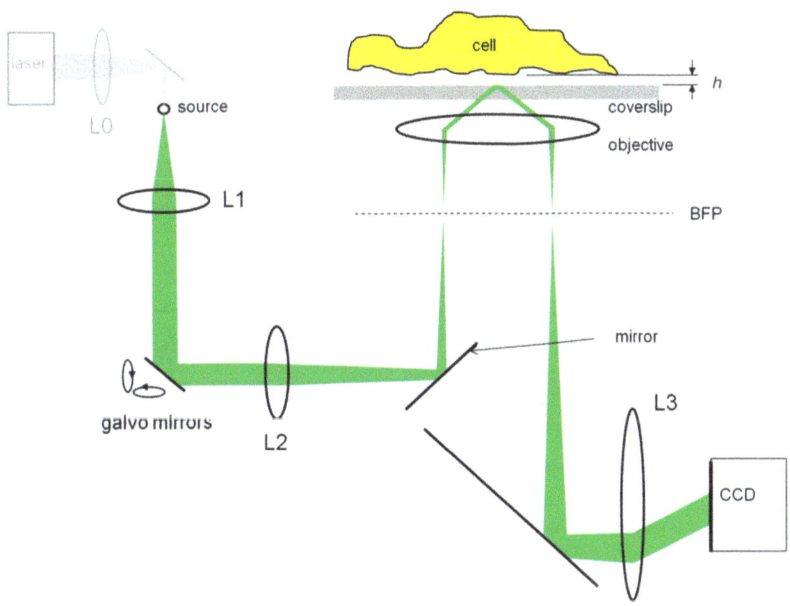

Figure 6.1. Optical configuration used for pseudoimaging of sample refractive indices. The light source can be either a small light-emitting diode (LED) or the focal point of a laser beam. Lens L1 collimates the source. The galvo mirrors are placed close to a plane complementary to the sample plane. Lens L2 refocuses the beam to the back focal plane (BFP) of the high-aperture (NA 1.49) objective at an off-axis radial position, adjustable by programming the galvo mirrors. A 100% reflecting mirror (in place of the standard dichroic mirror used for fluorescence) is positioned across 15%–30% of the aperture to direct the incident light up toward the sample. Lens L3 recollimates the reflected beam and casts it upon a CCD array positioned at a plane complementary to the sample plane. The distance scales are adjusted for pictorial clarity. Reprinted from Bohannon *et al* (2017), copyright Microscopy Society of America 2017.

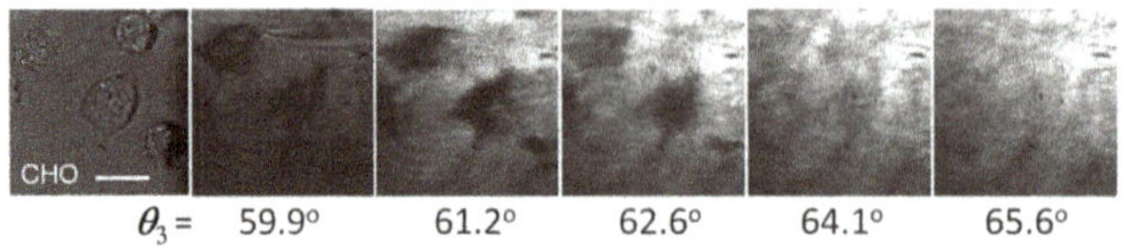

Figure 6.2. Raw reflected light images from an image sequence of CHO cells with selected increasing θ_3. A transmitted light view at the extreme left shows where the cells are located. The images are cropped to 800 × 800 pixels from the original 2 K × 2 K. Scale bar = 20 μm. Reprinted from Bohannon *et al* (2017), copyright Microscopy Society of America 2017.

the resolution of the image at any incidence angle is somewhat reduced, to ~1 μm rather than the ~0.2 μm that might be expected had the full aperture been used.

Figure 6.2 shows actual reflection images of a sample of Chinese hamster ovary (CHO) cells in culture, for a range of incidence angles. In the leftmost panel, almost all of the sample is subcritical, with a reflection of much less than 100%. In the rightmost panel, most of the sample is supercritical, with a reflection close to 100%. The images are very sensitive to small changes in incidence angle, corresponding to the steepness of the curves in figure 3.6.

By analyzing the intensity vs θ_3 curve for each location on the sample and then fitting the results to the theory given by equation (3.19), the values of the refractive index (and the spacing h between the coverslip and the bottom of the cell) can be deduced for each location. With these results, spatial maps of n_1 and h can be constructed (figure 6.3).

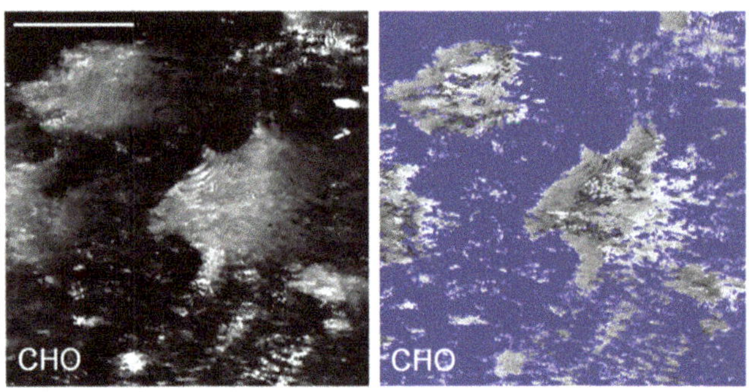

Figure 6.3. Calculated spatial maps for n_1 and h, constructed from a pixel-by-pixel analysis of each I vs θ_3 curve for the stack from which the sequence shown in figure 6.2 was sampled. The grey scale range for n_1 is from 1.330 to 1.390. All values of n_1 less than 1.330 are shown as black; these can be as low as 1.32 due to noisy I vs θ_3 curves. The corresponding spatial maps for h have a grey scale ranging from 10 to 290 nm. In the off-cell regions, where n_1 is generally between 1.33 and 1.35, theory predicts that h does not strongly affect the I vs θ_3 curves. The determination of h in those regions is consequently ill-defined or noisy at best. Those regions are depicted in a uniform blue. Scale bar = 20 μm. Reprinted from Bohannon *et al* (2017), copyright Microscopy Society of America 2017.

6.2 Axial position and motion of cell components

The determination of the z distances between a single fluorophore and the TIR surface was discussed in section 2.3 (ignoring the effects of near-field emission) and in sections 4.6 and 4.13 (including near-field effects). TIR can also view the axial motions of supramolecular assemblies in cells, such as the plasma membrane or secretory granules. The examples given here are intended not as a complete biological description, but only as illustrations of the techniques.

By recording a series of TIR images of a fluorescent plasma membrane-labeled cell over time, local changes in intensity can be tracked and analyzed. For example (figure 6.4), a series of images (time interval of 0.1 s) of a chromaffin cell (labeled with lyn-citrine) shows subtle localized intensity variations ('speckles,' extending in size to ~1 μm). These variations change from one frame to the next. Some regions get brighter, and others get dimmer. These differences can be displayed more vividly by creating a running difference stack using successive image subtractions; i.e. (frame 1 – frame 0), (frame 2 – frame 1), etc. These frame-to-frame differences are largest just inside the periphery of the cell.

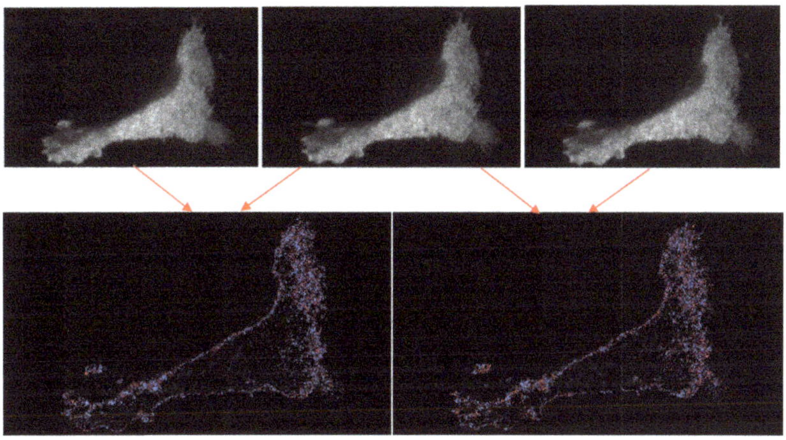

Figure 6.4. (Top row) Three successive frames of a chromaffin cell labeled with lyn-citrine, illuminated by TIR, with a time interval of 0.1 s from left to right. (Bottom row) Difference images as indicated by the arrows. Color coding: the differences range from very negative (deep blue) to zero (black) to very positive (bright red).

The question is: are the intensity variations over space and time due to the frame-to-frame lateral (x–y) reshuffling of fluorophore clusters, or are they due to changes in the local z positions of the membrane (i.e. bulgings and infoldings)? If the latter is the case, the relative size of the fluctuations in a chosen region of interest should appear much larger in TIR than under epi-illumination (EPI, a common abbreviation for subcritical angle illumination). In the chromaffin cell example here, this is indeed the case, by about a factor of ten. (The fluctuations are substantial, ranging in amplitude up to about 0.6 of the mean TIRF intensity of the cell.) Therefore, the TIR intensity variations around the cell periphery appear to be due to the local membrane 'flickering' up and down in the z direction.

To infer a z position (possibly time-dependent) from the observed intensity can be complicated because the near field of the emission forces a departure from the pure exponential decay of the TIR excitation (see chapter 4). However, assuming the decay is a pure exponential decay in z (perhaps augmented by some z-independent constant representing scattering) makes the analysis much simpler, and so that is what we assume in this section.

If the fluorophore is a submicroscopic point, then the conversion of intensity to z position is straightforward (figure 6.5).

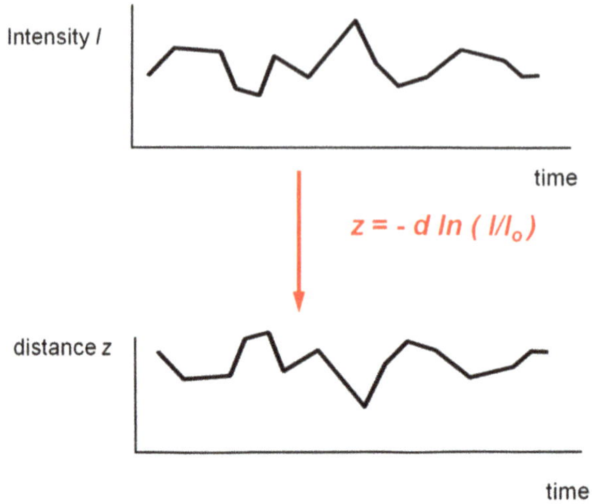

Figure 6.5. Conversion of intensities to z positions, with the assumption that the evanescent field is a pure exponential with no scattering and ignoring near-field emission effects.

However, cellular organelles are not submicroscopic points; they may have irregular shapes and dimensions comparable to or larger than the evanescent field, so that they span a range of evanescent intensities (figure 6.6). Nonetheless, the simple conversion depicted in figure 6.5 is still valid. If an irregularly shaped fluorescent object moves up, i.e. away from the TIR surface, every position in the object moves up by that amount, and the fluorescence of every position decreases by a factor of $\exp(-\Delta z/d)$. Therefore, the whole object, regardless of size or shape, also decreases its total fluorescence by that same factor.

This conclusion is, of course, only valid if the object does not change its shape or orientation in the course of its shift in z.

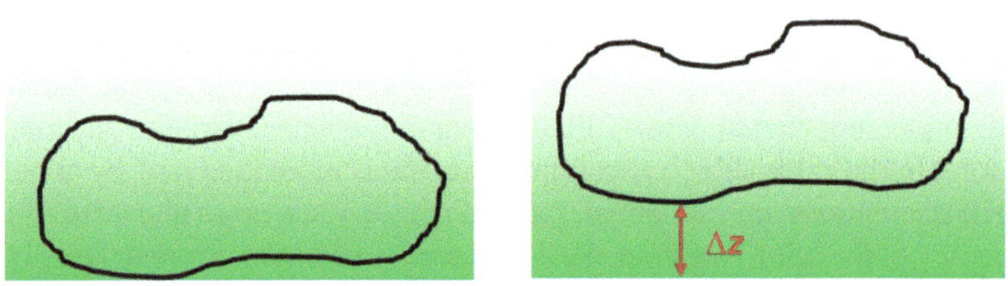

Figure 6.6. Schematic view of an irregular large object shifting upward in the evanescent field.

If the object is spherical, possible rotations do not matter. This is the case for secretory granules in the cytoplasm of bovine chromaffin cells. Their diameter is approximately 300 nm, somewhat larger than a typical evanescent field depth (50–100 nm). Many of those granules are near the plasma membrane, but not all of them. They occupy a range of z positions, and most of them also 'jiggle' in time in all three dimensions.

The amount of movement between successive frames is much less than the diameter of a granule, but it is easily detectable. In the z direction, a motion of as little as one tenth of the evanescent field depth results in a measurable change of intensity. In the x and y directions, motions down to about much less than the resolution of the microscope (~200 nm) can be measured by tracking the centroid of the image of each granule. Figure 6.7 shows TIR images of secretory granules at two successive time points separated by 0.1 s, followed by a difference image. Any deviation from grey represents a change of intensity: black represents less intensity and white represents more. A white region immediately adjacent to a black region is caused by the motion of a granule in the direction from black to white.

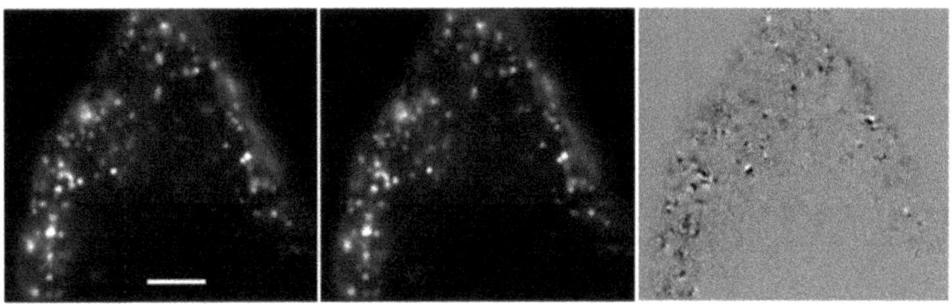

Figure 6.7. Successive TIR images of plasma membrane-proximal secretory granules in bovine chromaffin cells. The right panel is a difference image of the middle image minus the left image. The granules are transfected with a green fluorescent protein (GFP)-type (cerulean)-neuropeptide Y protein complex that the cell incorporates into the granules. Space bar = 4 μm.

The absolute brightness of a granule is a measure of the z distance of the granule from the plasma membrane nearest to the TIR surface. (Without further investigations, the distance from the plasma membrane to the TIR surface is not well known.) Changes in the frame-to-frame brightness of a granule can be interpreted as a measure of the one-dimensional Δz distance through which the granule has traveled. This distance is related to the apparent diffusion coefficient D_z by

$$\langle(\Delta z)^2\rangle = 2D_z T, \tag{6.1}$$

where T is the time interval between successive frames, and the brackets $\langle\rangle$ refer to an average over either many granules in a similar environment, or over time (as long as the granule remains in a similar environment). In the lateral (x and y) dimensions, the relevant displacement ΔR (by the Pythagorean theorem) is:

$$(\Delta R)^2 = (\Delta x)^2 + (\Delta y)^2, \tag{6.2}$$

which (assuming the motion is blind to any distinction between x and y) is then related to the apparent lateral diffusion coefficient D_R by:

$$\langle(\Delta R)^2\rangle = 4D_R T. \tag{6.3}$$

Measuring the dependence of the diffusion rate on the z position is possible only because of the uniquely steep gradient of TIR illumination in the z direction (see Allersma *et al* 2006 for more details). In general, figure 6.8 shows that both $\langle(\Delta z)^2\rangle$ and $\langle(\Delta R)^2\rangle$ are significant and that they increase with z. Presumably, there is some structure or bonding near the surface that slows them down.

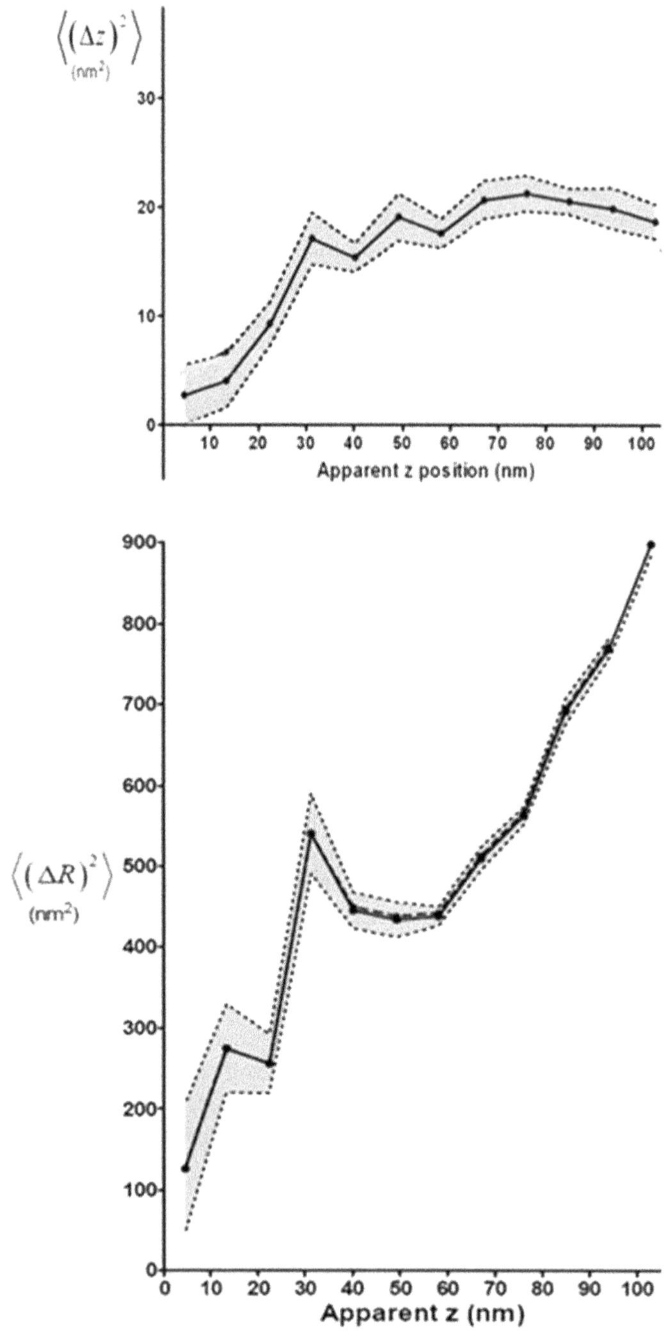

Figure 6.8. $\langle(\Delta z)^2\rangle$ and $\langle(\Delta R)^2\rangle$ versus z for membrane-proximal secretory granules in bovine chromaffin cells. The apparent z position is based on the observed average intensity between successive pairs of z or R measurements of a particular granule. The brightest granule in the cell is used to define $z = 0$. The gray band represents the 95% confidence range values. Note the difference in ordinate scales: $\langle(\Delta R)^2\rangle$ is much larger than $\langle(\Delta z)^2\rangle$. Adapted with permission from Allersma M W *et al* (2006)

The association of $\langle(\Delta z)^2\rangle$ and $\langle(\Delta R)^2\rangle$ with diffusion coefficients of course assumes that the process really is diffusive, i.e. a random walk. It turns out that this is not exactly the case. In a real diffusive process, $\langle(\Delta z)^2\rangle$ and $\langle(\Delta R)^2\rangle$ should be linearly proportional to time. However, the granules appear to be limited or entrapped somehow, at least in the z direction: they do not travel as far over long times as would be expected based on their travel over shorter time periods (see figure 6.9)

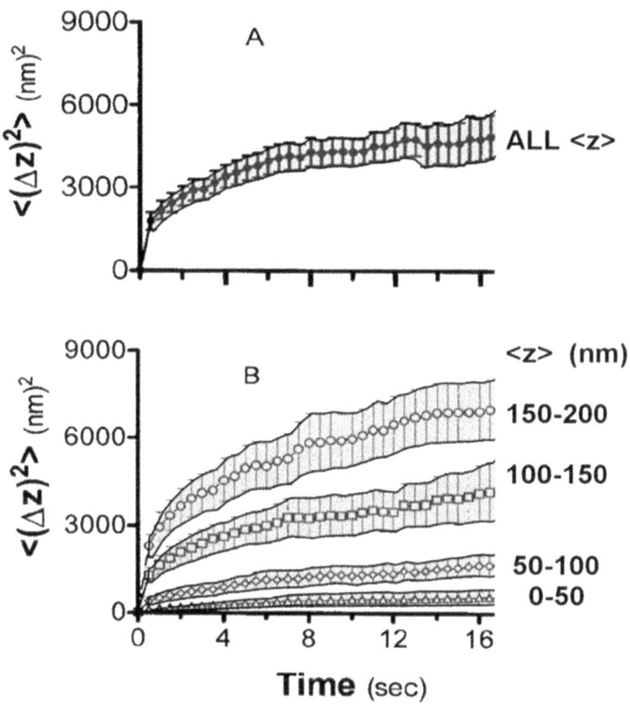

Figure 6.9. Measured travel $\langle(\Delta z)^2\rangle$ over long time periods. The concave-down shapes indicate restricted motion. Nonetheless, granules farther from the plasma membrane do travel farther.

In all measurements of z or Δz, care must be taken to: (a) subtract the estimated background fluorescence due to TIR scattering, which typically may be 10% to 20% (see section 3.5); and (b) account for intensity noise sources other than fluctuations in the z position, such as shot noise and fluctuations in the camera's dark count. Allersma *et al* (2006) suggest methods for taking these factors into account.

Evanescent field excitation enables several biologically surprising observations to be made regarding granule motion and the secretion event itself: (a) not all granules are secreted when the cell is stimulated, but the ones that are secreted (an event which appears as a bright flash that quickly dissipates) tend to be the newcomers and fast movers, not the ones which were long parked in close apposition to the membrane; (b) granules tend to secrete at certain preferred locations, not just anywhere; (c) proteins in the granule membrane tend to disperse into the plasma membrane upon fusion of the granule with it; (d) a typical granule tends to speed up just before fusion, while moving into a fresh region that it has not already explored; (e) adenosine triphosphate (ATP) removal slows granule motion; (f) granules tend to be transiently tethered, then make large jumps to new areas.

A variation of TIRF (called TIR aqueous fluorescence (TIRAF)) used to identify cell–substrate contact regions involves doping the solution surrounding the cells with a nonadsorbing and impermeable fluorescent volume marker; focal contacts then appear relatively dark (figure 6.10(a)) where the fluorescent dye layer is very thin. The unique advantage of this approach is that it reports a local intensity that is a weighted integral over the extracellular distance between the TIR substrate and the plasma membrane, which can be converted into an absolute local distance separating the surface from the cell. The fluorescence is, however, not chemically specific for any biochemical component.

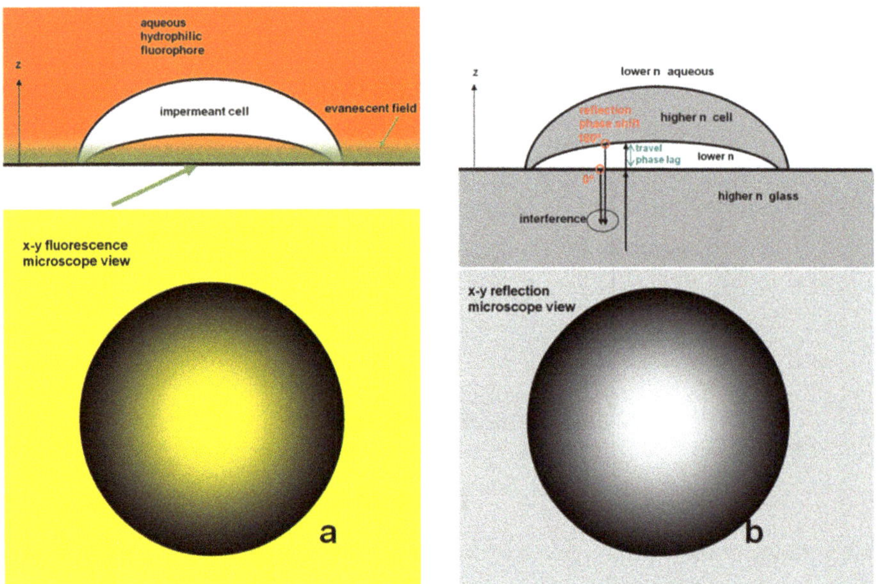

Figure 6.10. Schematic diagrams of a cell, comparing (a) TIRAF (yellow fluorescence) and (b) interference reflection contrast (IRC).

A related technique which also can provide information about cell–substrate spacing does not involve TIR or fluorescence at all, although it does share the word 'reflection' in its name: interference reflection contrast (IRC) (figure 6.10(b)). The sample system can be viewed as three layers: glass substrate–water–cell. The incident light partially reflects from the substrate–water index without a phase shift (because the light there reflects from a higher-to-lower refractive index) and also partially reflects from the water–cell interface with a 180° phase shift (the light reflects from a lower-to-higher index). These two reflections can then interfere, since they are the same color and easily within a coherence length of each other. If the water layer approaches zero thickness, the two reflections will be 180° out of phase, and the combined reflection (as seen through the objective) appears relatively dark. Despite the distinction between IRC and TIRF, the two techniques can be used on the same sample. The IRC intensity is simultaneously sensitive to the cell–substrate spacing and to the local refractive index of the cell. TIRF is sensitive to the cell–substrate spacing and to the local concentration and accessibility of fluorophore molecules. By comparing images obtained using the two techniques, ambiguities can be resolved.

6.3 Quenching with a metal film

The z dependence of the evanescent field can be radically altered from its exponential form by localized fluorescence quenching near the TIR surface. One approach uses a metal film on the surface, as depicted graphically in figure 4.12(c) for a single dipole emitter. Within the first 10 nm of the z position, fluorescence from the emitter is strongly quenched; it then reaches a maximum at ~130 nm and then decays toward zero in a roughly exponential fashion (with the expected evanescent decay constant for the incidence angle chosen) at larger z positions. This effect can be utilized with biological cells. For example, a fluorophore that reversibly (and weakly) binds to a cell may also bind nonspecifically to the substrate upon which the cells are adhered. This unwanted background can overwhelm the weak but specific binding of interest. A metal film can quench that background while preserving the fluorescence bound to cells more distant from the substrate (see figure 6.11).

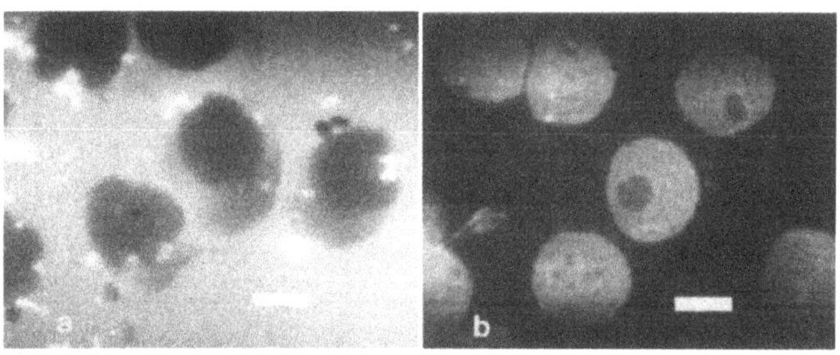

Figure 6.11. Quenching of near-substrate TIR fluorescence by an aluminum film coated (~20 nm thick) on the substrate. Panel (a) shows the bare glass case; panel (b) shows the metal film case. In both cases, the surface was pre-coated with poly-lysine, which leads to strong adherence of cells. Here, erythrocytes were adhered then split open by osmotic shock, becoming flattened and split open by strong adherence to the surface. The erythrocyte ghosts are in equilibrium with a fluorescein-labeled epidermal growth factor, and the TIR is prism-based. Space bar = 5 μm. Reprinted from Fulbright *et al* (1993), copyright (1993) Plenum Publishing Corporation, with permission of Springer.

Another approach to quenching is actually a variation of stimulated emission depletion (STED), adapted to TIRF illumination. A pulsed TIRF field excites fluorophores in the usual exponential decay pattern. Another TIR pulse, following immediately at a second wavelength, stimulates emission from its brightest zone (near the surface), thereby quenching it from future probe pulses while preserving the prior excitation of fluorophores at larger z. The result is a relatively thin optical slice 'floating' at a region of larger z. (In standard EPI-STED, a brief pulse of a donut-shaped focused spot stimulates fluorescent emission in a small ring of already-excited fluorophores, but not at its center.) The displacement of the fluorophore population from the TIR surface depends on the angles of incidence of the excitation and quenching beams and the optical power of the quenching beam.

6.4 Image sharpening in TIR

In standard EPI-illumination fluorescence imaging, a significant improvement in the (otherwise) rather poor axial resolution can be obtained by post-acquisition image deconvolution. This approach (actually, a range of approaches) involves acquiring a stack of images through a range of focal positions. In an iterative mathematical process, possibly (but not necessarily) utilizing knowledge of the 3D point spread function (PSF), the information from each focal plane is then used to correct the image of its neighbors above and below. The results are impressive but suffer from a couple of problems. One is that the final result is not necessarily linear: i.e. doubling the original object intensity from one location to another does not necessarily lead to intensity doubling in the deconvolved image stack. Another problem is that the computations take a considerable time.

Sharpening TIR images is much faster, because only one focal plane is present and all the sharpening takes place in the x and y directions. The procedure is basically 'high-pass' filtering, the characteristics of which are adjusted to correspond to the inherent blurring by the PSF. Figure 6.12 shows an example of the effect.

The simplest approach is as follows. Let $f(\mathbf{r})$ *be* the 'true' 2D image that would be formed by a lens from an object in **r**-space if the PSF were a perfect δ function (a physical impossibility, of course). The function $f(\mathbf{r})$ is what we wish to calculate. Let $g(\mathbf{r}')$ be the actual 2D TIRF image in image space because of the finite size of the PSF. Let $h(\mathbf{r}')$ be the 2D PSF, where $\mathbf{r}'$ is measured in image space from the center of the image of a typical object point, wherever it is in the object. Then:

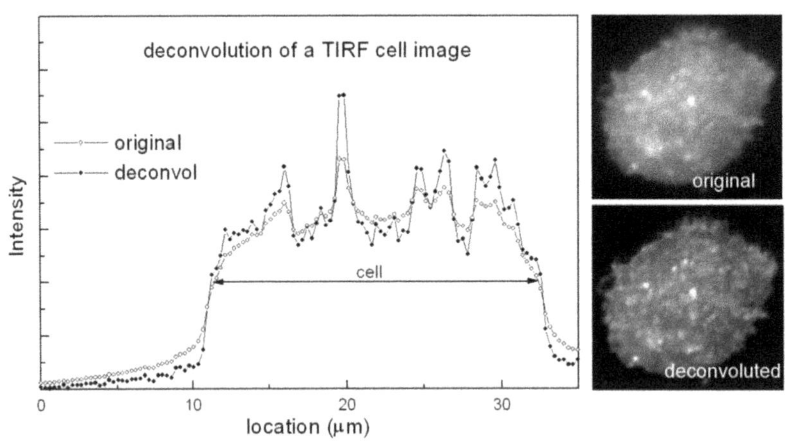

Figure 6.12. Image sharpening of a (2D) TIR image by deconvolution with the known PSF.

$$g(\mathbf{r}') = \int h(\mathbf{r}'')f(\mathbf{r}' - \mathbf{r}'')d^2\mathbf{r}''. \tag{6.4}$$

This integral is a 'convolution,' denoted for brevity by:

$$g = h*f. \tag{6.5}$$

A two-dimensional Fourier transform of both sides of equation (6.4) gives:

$$\tilde{g}(\mathbf{k}) \equiv \int g(\mathbf{r}')e^{-i\mathbf{k}\cdot\mathbf{r}'}d^2\mathbf{r}' = \int h(\mathbf{r}'')d^2\mathbf{r}''f(\mathbf{r}' - \mathbf{r}'')e^{-i\mathbf{k}\cdot\mathbf{r}'}d^2\mathbf{r}', \tag{6.6}$$

where the superscript tilde '~' is shorthand notation for a Fourier transform.

Let $\mathbf{r} \equiv \mathbf{r}' - \mathbf{r}''$, so for a fixed $\mathbf{r}''$, $d^3\mathbf{r}' = d^3\mathbf{r}$. Equation (6.3) then becomes:

$$\tilde{g}(\mathbf{k}) = \int h(\mathbf{r}'')e^{-i\mathbf{k}\cdot\mathbf{r}''}d^2\mathbf{r}'' \int f(\mathbf{r})e^{-i\mathbf{k}\cdot\mathbf{r}}d^2\mathbf{r} = \tilde{h}(\mathbf{k})\tilde{f}(\mathbf{k}). \tag{6.7}$$

Therefore, the Fourier transform of a convolution is the product of the Fourier transforms of the functions being convolved. (This result is known as the 'convolution theorem.') From equation (6.7), we get

$$\tilde{f}(\mathbf{k}) = \frac{\tilde{g}(\mathbf{k})}{\tilde{h}(\mathbf{k})}. \tag{6.8}$$

In principle, all we need to do is invert the Fourier transform to get back to $g(\mathbf{r})$, which was our goal:

$$f(\mathbf{r}) = \int \frac{\tilde{g}(\mathbf{k})}{\tilde{h}(\mathbf{k})}e^{+i\mathbf{k}\cdot\mathbf{r}}d^2\mathbf{r}. \tag{6.9}$$

The Fourier transform of the PSF, appearing here as $\tilde{h}(\mathbf{k})$, is also known as the 'contrast transfer function' (CTF) or the 'optical transfer function' (OTF). The CTF shows how well each spatial frequency (along each axis) survives the trip from object to image. High spatial frequencies are lost because the PSF is a blurry region rather than a sharp point. A crude approximation of the PSF could be the Airy disk. Alternatively, the PSF can be calculated as shown in section 4.10, or measured experimentally.

In practice, equation (6.9) must be handled with care, because for certain high $\mathbf{k}$, both the numerator and denominator in the integrand approach zero. If experimental data is used for $\tilde{g}(\mathbf{k})$ and $\tilde{h}(\mathbf{k})$, their ratio is very noisy.

6.5 Polarized excitation TIRF

As discussed in section 2.2 and shown in figure 2.1, supercritical p-polarized incident light produces a TIR evanescent field with a unique polarization: predominantly in the z direction (the optical axis of the microscope, usually the normal to the sample substrate). Standard EPI illumination (for which the incidence angle is close to zero) produces only x and y polarizations; see figure 6.13.

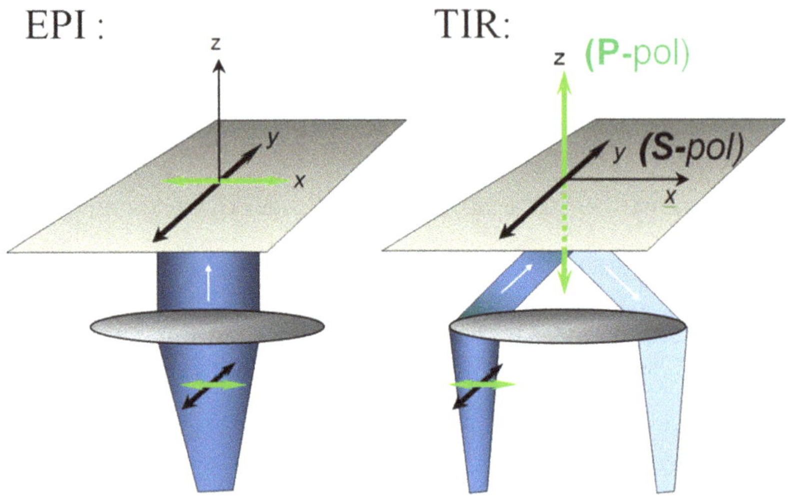

Figure 6.13. The two polarizations of incident light and the polarizations they produce at the sample plane, for EPI and TIR incidence. Incident light in p-pol is shown in green; s-pol is shown in black.

This unique polarization of the p-pol evanescent field can be used to detect submicroscopic orientational nonuniformities, such as indentations or bulges in plasma membranes, features that may not be visible using other optical techniques. This technique depends on the fluorophore being highly oriented in its target. For example, diI is highly oriented such that its absorption and emission dipole moments are close to parallel to a membrane surface (Axelrod 1979). Figure 6.14 shows how this can be applied to membrane orientation.

In general, TIRF images show a convolved mix of local fluorophore orientations with concentrations and distances from the substrate. What combination of p-pol

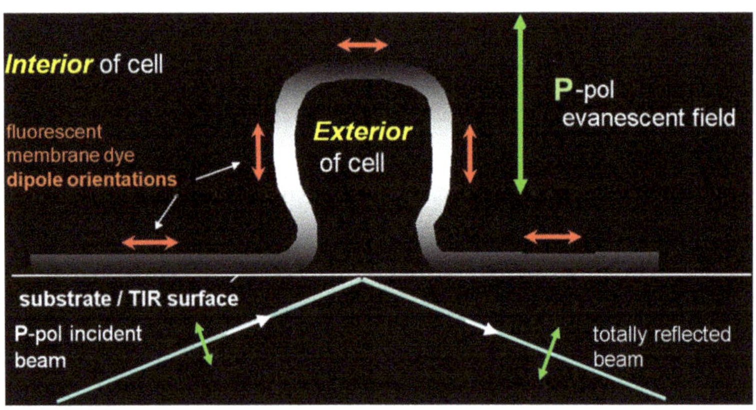

Figure 6.14. Polarized excitation TIRF (when used with a diI membrane fluorophore) highlights indented regions of the membrane (such as endo- or exocytosis sites). In regions where the membrane (and the diI) have orientations that contain a z component, the p-pol evanescent field successfully excites the fluorophore. In other places, where the membrane is parallel to the substrate, the diI is not excited.

and s-pol excited images reports purely orientational distributions, and what combination reports purely concentration and distance?

To calculate the answer, we set up a polar coordinate system: the sample plane (at which TIR occurs) is the x–y plane, and the optical axis of the microscope is the z axis (see figure 6.13, right panel). The TIR evanescent field propagates in the x–y plane along x, s-pol excitation is polarized along y, and p-pol excitation is polarized along z. (Actually, a small amount of p-pol intensity is polarized in the x direction, depending on the incidence angle of the TIR beam, as discussed in section 2.2, but we ignore it here for simplicity). In this coordinate system, θ and ϕ are the polar angle (measured from the z axis) and the azimuthal angle (measured around the z axis), respectively. For a particular single fluorophore whose absorption dipole is oriented at (θ,ϕ), the probability of excitation caused by p-pol or s-pol light is proportional to the square of the component of the dipole along the z or x axes, respectively. These square components are proportional to $\cos^2\theta$ and $\sin^2\theta\cos^2\phi$, respectively.

Given that the fluorophore becomes excited, the emission can be resolved into two components, one along the z axis (i.e. perpendicular to the substrate) and the other in the x–y plane (i.e. parallel to the substrate). We assume here for simplicity (and because it is usually approximately true), that the emission dipole and the absorption dipole are parallel to each other. The squares of each of these components, which are proportional to $\cos^2\theta$ and $\sin^2\theta$, each contribute to the emitted light collected by the objective with efficiencies of $Q^{\perp}$ and $Q^{\parallel}$ (see section 4.6), respectively.

An actual sample, of course, contains numerous fluorophores whose dipoles possibly point in different directions. We assume that the normalized orientational distribution of dipoles is $\eta(\theta,\phi)$. The emission light intensities P and S gathered by an objective (and imaged at a single pixel) resulting from p-pol and s-pol excitation, respectively, are

$$
\begin{aligned}
P &= \int C\eta(\theta,\phi)\cos^2\theta(Q^{\perp}\cos^2\theta + Q^{\parallel}\sin^2\theta)\sin\theta d\theta d\phi \\
S &= \int C\eta(\theta,\phi)\sin^2\theta\cos^2\phi(Q^{\perp}\cos^2\theta + Q^{\parallel}\sin^2\theta)\sin\theta d\theta d\phi,
\end{aligned}
\tag{6.10}
$$

where C is a number proportional to the local effective total amount of fluorophore at the pixel under view. C implicitly includes the decay effect of the evanescent field in the z direction, such that fluorophores at larger distances from the surface simply contribute less to C. C also implicitly includes the foreshortening effect of tangential viewing and all the multiplicative constants and efficiencies of absorption and emission that convert concentration into units of light intensity. The assumptions in equation (6.10) are that no polarizer is used in the emission system, the dichroic mirror introduces no polarization bias, the absorption and emission dipoles are parallel to each other, and that the orientational distributions are not a function of the z distance to the TIR substrate.

Clearly, the P/S ratio only reports on the orientational distribution $\eta(\theta,\phi)$, because the C factors cancel upon taking the ratio. The goal of the following derivation is to find the opposite: a combination of P and S that only reports on C and does not depend on $\eta(\theta,\phi)$.

All of the possible biases of the objective toward gathering light from one excited dipole orientation or another are contained in the factors $Q^{\perp}$ and $Q^{\parallel}$. For a very high aperture (NA $\geqslant$1.40), these factors are within 20% of each other (see the right panel of figure 4.6), so we will set them both equal to unity in equation (6.10). This leads to a collapse of the integrand factor involving $Q^{\perp}$ and $Q^{\parallel}$ to unity. Note that this step is only valid for 1.40, 1.45, and 1.49 NA objectives, and even then, it is just an approximation. The normalization requirement is that

$$\int \eta(\theta,\phi) \sin\theta d\theta d\phi = 1. \tag{6.11}$$

We now assume azimuthal symmetry in fluorophore orientation such that $\eta(\theta,\phi) = \eta(\theta)$; i.e. there is no particular orientation preference for ϕ in the x–y plane. This is expected to be the case in many biological organelles (plasma membrane, secretory granules, etc) if we gather light from a large enough area. The normalization (equation 6.11) then becomes

$$\int \eta(\theta,\phi) \sin\theta d\theta = \frac{1}{2\pi}. \tag{6.12}$$

Equation (6.10) now becomes

$$\begin{aligned} P &= 2\pi \int C\eta(\theta) \cos^2\theta \sin\theta d\theta \\ S &= \pi \int C\eta(\theta) \sin^2\theta \sin\theta d\theta = \pi \int C\eta(\theta)(1 - \cos^2\theta) \sin\theta d\theta \\ &= \pi C \int \eta(\theta) \sin\theta d\theta - \frac{P}{2} = \frac{C - P}{2}. \end{aligned} \tag{6.13}$$

Therefore,

$$P + 2S = C. \tag{6.14}$$

This expression is true, regardless of the exact form of the orientational distribution $\eta(\theta)$ (as long as it is azimuthally symmetric). The sum P+$2S$ is proportional only to the effective concentration. As long as we view with a very large-aperture objective and bin together a large enough image area to ensure that the orientational distribution within the area is azimuthally symmetric, P+$2S$ is approximately proportional only to the local concentration convolved with the evanescent exponential decay, and invariant with respect to the details of the azimuthally symmetric orientation. Because $Q^{\perp}/Q^{\parallel}$ undulates as a function of z (figure 4.6), there is no simple algebraic combination of P and S that is exactly invariant to the orientation for all geometrical shapes of the membrane.

As examples, consider two particular azimuthally symmetric $\eta(\theta)$ functions: (a) the fluorophore diI in a planar plasma membrane parallel to the substrate and (b) a spherical granule much smaller than the size over which P and S are evaluated.

We assume for now that diI intercalates into lipid bilayer membranes with its dipole moments parallel to the local membrane surface (Axelrod 1979). The $\eta(\theta)$ functions are (in properly normalized form):

$$\eta_{\text{plane}} = \frac{1}{2\pi}\delta(\cos\theta) \qquad \eta_{\text{sphere}} = \frac{1}{4\pi}. \tag{6.15}$$

From equation (6.13), we then see that the P/S ratio provides a very sensitive test of local orientation:

$$\left(\frac{P}{S}\right)_{\text{plane}} = 0 \qquad \left(\frac{P}{S}\right)_{\text{sphere}} = 1. \tag{6.16}$$

For the same C (and hence the same surface area) for both the plane and the sphere, $P+2S$ remains constant, as expected from equation (6.14).

This idealized calculation must be altered to include numerous real effects: finite evanescent field depth, structures deviating from perfect planes or spheres, pixelation, and sphere sizes on the edge of optical resolvability. For details, see Anantharam *et al* (2010).

The polarized excitation TIRF effect can be vividly seen in biological samples with known shape features, for example, an erythrocyte (red blood cell) and a macrophage (figure 6.15).

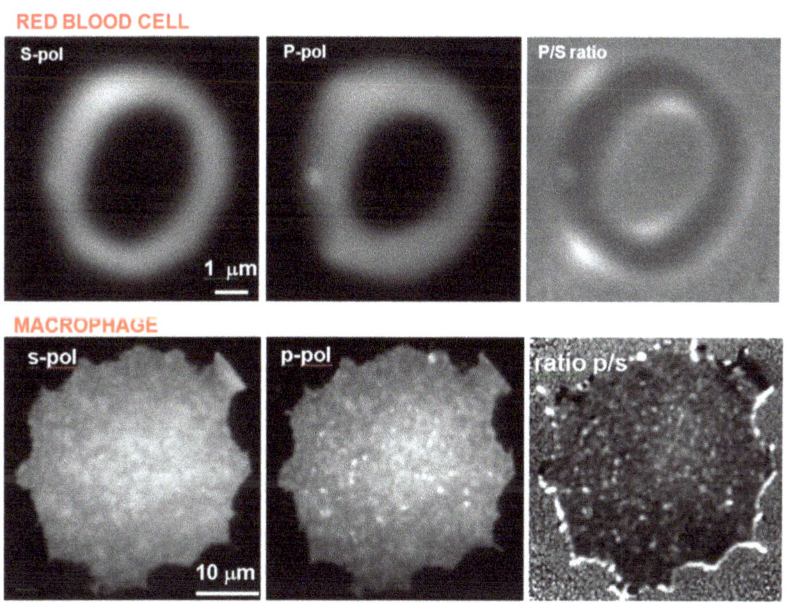

Figure 6.15. Polarized excitation TIRF of a red blood cell (top) and a macrophage (bottom); the plasma membrane is labeled with diI. The P/S ratio highlights the cell edges, where the membrane slants upward from the substrate. In the case of macrophages, the P/S ratio also highlights the highly irregular surface. Videos show that these irregularities are highly dynamic. Reprinted from Sund *et al* (1999), copyright (1999), with permission from Elsevier.

Unexpected features are also revealed by the *P/S* ratio, illustrated here (figure 6.16) for a small region of a chromaffin cell during the process of secretion, showing a membrane dimpling at the site of the secretion, which lasts for >0.4 s after the secretion event.

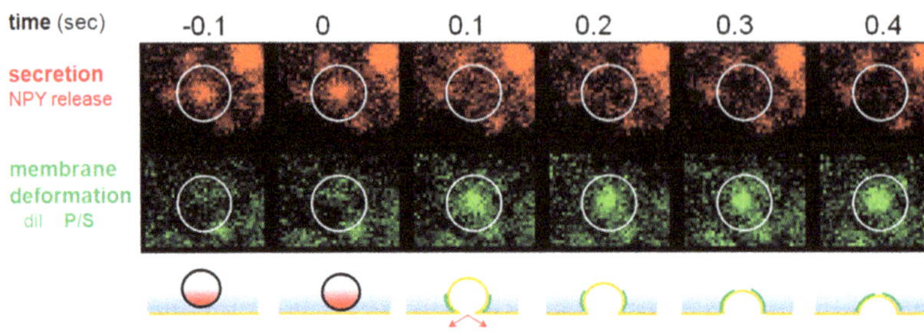

Figure 6.16. A chromaffin cell granule secretion event causes a long-lasting subsequent indentation in the plasma membrane. The camera takes a repeating set of three successive images: a short wavelength excitation to visualize the GFP-marked neuropeptide Y (NPY) inside the granule, and then a longer-wavelength excitation in the p and then s polarizations to visualize diI and its orientation. The circles have a 1.2 μm diameter. The secretion event is characterized by the sudden loss of NPY, and the *P/S* ratio shows the subsequent membrane dimpling. The bottom row is an inference about the concurrent structural events. Adapted with permission from Anantharam *et al* (2011).

6.6 Variable-depth TIRF

In some systems, the concentration $C(z)$ near the surface is not necessarily a 2D planar structure (presumably a delta function profile at $z = 0$) but instead has an interesting z dependence, even within the thin evanescent field depth.. By varying the depth d of the evanescent field, some information about $C(z)$ can be obtained. Qualitatively, this is simple: a deeper evanescent field excites fluorophores at two different closely spaced z' levels more equally, whereas a shorter evanescent field more strongly favors the fluorophore at the smaller z'. Quantitatively, the problems of calculating $C(z')$ have been discussed in section 4.6.3. For a $C(z')$ that is orientationally isotropic, the fluorescence can be written as a function of depth d as follows (as adapted from equation (IV.32)):

$$F(d) = \xi(\theta, \lambda) \int_{z'=0}^{\infty} C(z')Q(z')e^{-z'/d}dz' = \xi(\theta, \lambda)L[C(z')Q(z')]. \tag{6.17}$$

where L represents the Laplace transform of $C(z')Q(z')$. Taking the inverse Laplace transform L^{-1} of both sides gives:

$$L^{-1}[F(d)] = \xi(\theta, \lambda)C(z')Q(z') \tag{6.18}$$

so

$$C(z') = \frac{L^{-1}[F(d)]}{\xi(\theta, \lambda)Q(z')}. \tag{6.19}$$

$Q(z')$, the collection efficiency of the objective, is a complicated function that can be calculated numerically, but at least it is known. $F(d)$ can be experimentally measured over a range of d (as many measurements as possible in order to calculate a smooth inverse Laplace transform). The depth d can be varied by either changing the incidence angle θ or the incident light wavelength λ (see equation 2.12). Experimental attention then turns to the prefactor $\xi(\theta, \lambda)$. Apart from various constants, $\xi(\theta, \lambda)$ contains the intensity of the evanescent light at $z' = 0$, which is a strong function of the incidence angle and the wavelength. This dependence must be incorporated from theory, or experimentally measured for the optical system being used, before $C(z')$ can be reported accurately.

6.7 Optical force in an evanescent field

A strong light intensity gradient, such as that produced by a focused laser beam, or an evanescent field, can generate a force on a microscopic object embedded in the gradient. Although the effect may be coordinated with fluorescence observation, it does not necessarily involve fluorescence at all. The purpose of nonfluorescent 'optical tweezers' is to apply optical force to hold or move microscopic or submicroscopic particles. This purpose has several specific applications in biophysics, including bringing two reactants together (either or both of which may be attached to the surface of a microscopic latex bead); to measure the force between reactants (because the force can be calibrated); and to elongate or stretch some microscopic structure such as a cytoskeletal filament. In TIRF, the fluorescence aspect is used to visualize the object, but the force created by the evanescent field may act to pull mobile objects (such as cellular organelles) closer to the TIR substrate.

For an intuitive idea of the origin of this force, we model the object as a dielectric sphere, with an index of refraction greater than that of its surroundings, and in fact, great enough to convert the evanescent light into propagating light. In the large ($\gg\lambda$) particle limit, the effective force can be understood qualitatively from the viewpoint of geometrical (ray) optics, in which refraction and reflections at the bead surfaces impart momentum from the light beam (photons) to the bead (see figure 6.17). The brighter ray (the one nearer the TIR surface) refracts with an upward component to its momentum and thereby imparts a downward momentum component to the bead. The dimmer ray farther from the surface, in similar but opposite manner, delivers a downward momentum component to the bead. But because of the difference in

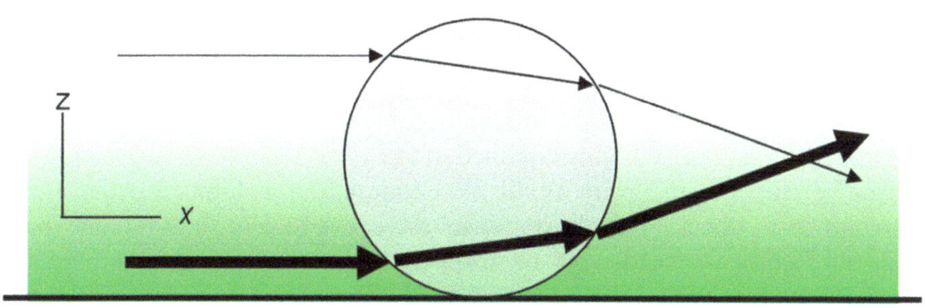

Figure 6.17. Ray model of momentum transfer from an evanescent field creating an optical force on a large bead.

intensity between the two rays, the downward momentum of the bead dominates, pushing it toward the brighter illumination.

This explanation is obviously crude. But a very elegant, exact, and far more general derivation that works for all sizes of spheres (but still spheres) was presented in Bekshaev *et al* (2013). (This is cited because it introduces a very powerful approach to the mathematics of evanescent fields by simply replacing the real propagation angle of a plane wave with a complex angle.) The result of that formulation is shown in figure 6.18.

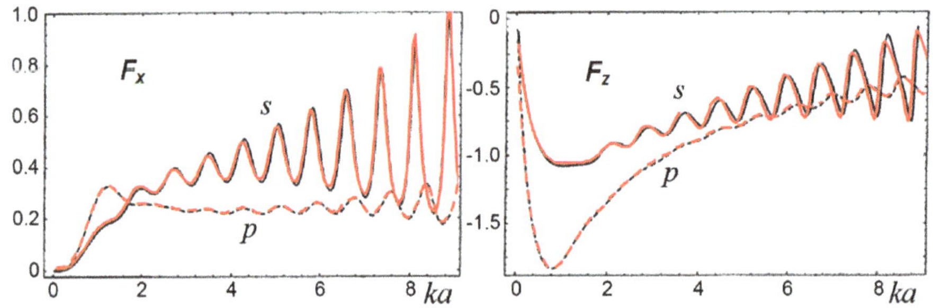

Figure 6.18. The optical force exerted upon a sphere of radius a (normalized to wavelength by $ka = 2\pi a/\lambda$), by an evanescent field produced with $n_3 = 1.75$, $n_1 = 1$, $n_{\text{sphere}} = 1.5$, and incidence angle $\theta_3 = 51°$. The force $\boldsymbol{F}$ has components in the x and z directions as defined in figure 6.17. The force is different for s-pol vs p-pol. $|\boldsymbol{F}|$ is normalized to the force exerted by an infinite propagating plane wave on the sphere, sometimes known as 'photon pressure.' Figure copied from Bekshaev *et al* (2013), copyright 2013 Optical Society of America.

As expected and seen from figure 6.18, the z component of the optical force is negative, meaning that it is directed toward the interface. The x component is 'downwind,' i.e. in the direction of travel of the evanescent field's wavefront.

The optical force effect conceivably could be used shove an organelle toward the substrate from its normal position, and then monitor its relaxation back upon reduction of the field, all while monitoring its position with TIRF. From this data, the viscosity, plasticity, and elasticity inside the cytoplasm could become manifest. Forcibly moving secretory granules toward the plasma membrane may alter their normal Brownian-like motion and possibly affect the probability of secretion.

Note that the optical force effect only occurs if the object has a refractive index different from that of its surroundings; it may be necessary to inject extrinsic plastic or glass microspheres of known characteristics into the cell to monitor mechano-elastic parameters quantitatively.

With the possibilities (discussed in section 5.11.5) for creating a single z-decaying evanescent field that is also highly confined in the x–y directions, a single molecule or bead can be trapped in a very small 3D illumination volume. Furthermore, by creating an x–y 'checkerboard' standing wave evanescent interference pattern, multiple molecules or beads can be trapped simultaneously, each in their own separate 3D microvolume. This scheme would allow viewing of chemical dynamics in many optical traps at the same time in one field-of-view, leading to a great enhancement of data throughput.

6.8 TIR/FCS and TIR/FRAP

There are numerous applications that combine TIR with fluorescence correlation spectroscopy (FCS) and with fluorescence recovery after photobleaching (FRAP), each with distinct setups and analyses. Here, we discuss the measurement of: (a) chemical kinetic rates of binding and unbinding from open volumes and 2D surface diffusion; (b) 3D molecular diffusion rates in open volumes; (c) molecular diffusion rates on or inside cellular organelles. FCS and FRAP in general are quite related in their mathematics but quite different in their experimental protocols; this remains true for their TIR variants.

6.8.1 Adsorption/desorption chemical kinetics

One visual cue for the transition from subcritical to supercritical illumination in a sample of fluorescent dye solution is the loss of the diffuse out-of-focus glow and its replacement by distinct fluorescent spots moving around and flickering at random. These fluctuations are directly due to single molecules (and small groups) engaging in Brownian motion through the evanescent field. The intensity fluctuations in small lateral regions (such as those defined by camera pixels or an image-plane optical diaphragm) can be monitored continuously in time. Their rapidity is proportional to the diffusion coefficient and the relative size of the fluctuations is related to absolute concentration and the degree of clustering or polymerization.

More generally, fluctuations can also arise from reversible binding and unbinding of fluorescent molecules to the surface, and can thereby be connected to chemical kinetic rates of adsorption k_1 and desorption k_2, as well as the diffusion coefficient D along the surface. The binding process may be either nonspecific (such as blood proteins adsorbed to a solid/liquid interface) or specific (such as soluble ligands binding to a surface bound receptor or a soluble antibody binding to a surface bound antigen).

This section presents the theoretical basis by which TIRF techniques directly measure those rate parameters, all without disturbing the chemical equilibrium of the system. The ability to measure kinetic parameters while the sample remains in chemical equilibrium is remarkable: most other rate-measuring techniques (pressure jump, concentration jump, temperature jump, etc.) depend upon perturbing the system and then watching it relax back toward equilibrium. Another unique feature of TIR/FCS and TIR/FRAP is that no change is required in spectroscopic properties between bound and unbound states.

In typical experiments described in this subsection, most solute molecules within the evanescent field are those actually bound to the surface. Thus, the measured fluorescence is mainly due to the surface bound solute molecules which are within a well-defined surface area. Subsequent subsections describe other experimental situations: e.g. those in which most molecules in the evanescent field are free (such that the bulk diffusion coefficient can be measured); or in which the molecules of interest are bound to subcellular organelles on living cells (such that diffusion on their membranes or interiors or the motion of the whole organelle can be measured).

When cast in the proper mathematical form, the theoretical treatments of TIR/FRAP and TIR/FCS are identical. We consider molecules of bulk concentration

$A(\mathbf{r}, z, t)$ freely diffusing in solution and reacting with free binding sites of surface concentration $B(\mathbf{r}, t)$ to form fluorescent complexes of surface concentration $C(\mathbf{r}, t)$. The reaction is represented by the chemical equation:

$$A(\mathbf{r}, z, t) + B(\mathbf{r}, t) \underset{k_2}{\overset{k_1}{\rightleftarrows}} C(\mathbf{r}, t). \tag{6.20}$$

The vector r is the position on the surface measured from the center of the observation area, z is the perpendicular distance from the surface to a point in the solution, and t is the time. The parameters k_1 and k_2 are surface adsorption and desorption rate constants. The equilibrium constant of the reaction is κ, where

$$\kappa = \frac{\overline{C}}{\overline{A}\,\overline{B}} = \frac{k_1}{k_2}. \tag{6.21}$$

$\overline{C}$, $\overline{A}$, and$\overline{B}$ are equilibrium concentration values, and are independent of position (assuming surface homogeneity).

We assume that each molecule within the evanescent field is bound to a surface site. The evanescent intensity $I(\mathbf{r})$ at the surface at a camera pixel or image-plane aperture can be written as

$$I(\mathbf{r}) = I_0\zeta(\mathbf{r}), \tag{6.22}$$

where I_0 is the maximum intensity and $\zeta(\mathbf{r})$ is a unitless profile function with unity maximum amplitude. For typical TIR/FRAP experiments, $\zeta(\mathbf{r})$ is the profile formed by the focused laser beam. For typical TIR/FCS experiments, $\zeta(\mathbf{r})$ is the transmission function of the image-plane aperture or camera pixel, typically a rectangle. The measured fluorescence $F(t)$ is then

$$F(t) = QI_0 \int \zeta(\mathbf{r})\ C(\mathbf{r}, t) d^2\mathbf{r}, \tag{6.23}$$

where Q is a product of the efficiencies of excitation light absorption and fluorescence emission and detection.

For **TIR/FRAP**, we solve for the monotonically decreasing function

$$G_p(t) \equiv \overline{F} - F(t), \tag{6.24}$$

where $F(t)$ is the fluorescence after a photobleaching flash at $t = 0$ due to unbleached fluorophore within the observation region; $\overline{F}$ is the equilibrium prebleach fluorescence (and the fluorescence after complete recovery); and the subscript p denotes photobleaching. We define

$$\begin{aligned} \mathcal{A}_p(\mathbf{r}, z, t) &\equiv \overline{A} - A(\mathbf{r}, z, t) \\ \mathcal{C}_p(\mathbf{r}, z, t) &\equiv \overline{C} - C(\mathbf{r}, z, t), \end{aligned} \tag{6.25}$$

where $\mathcal{A}_p(\mathbf{r}, z, t)$ and $\mathcal{C}_p(\mathbf{r}, z, t)$ are the bulk and surface concentrations of unbleached fluorescent solute molecules after the photobleaching flash at $t = 0$. $\overline{A}$ and $\overline{C}$ are the equilibrium values of the bulk and surface *total* concentrations of solute molecules (unbleached plus bleached), respectively. Combining equations (6.23–6.25) yields

$$G_p(t) = QI_0 \int \zeta(\mathbf{r}) \mathcal{C}_p(\mathbf{r}, t) d^2\mathbf{r} \tag{6.26}$$

For **TIR/FCS**, we solve for the temporal autocorrelation function

$$G_c(t) \equiv \left\langle \delta F(t) \delta F(0) \right\rangle, \tag{6.27}$$

where the subscript c denotes correlation experiments, and

$$\delta F(t) \equiv F(t) - \overline{F}. \tag{6.28}$$

$\delta F(t)$ is the spontaneous fluctuation of fluorescence $F(t)$ at time t away from the mean fluorescence F. Time zero is arbitrary. The brackets represent a thermodynamic ensemble average.

The fluctuations of the three chemical species from their equilibrium concentrations are δA, δB, and δC. We define A_c and C_c to be the temporal correlation functions of δA at $(\mathbf{r}, z)$ and δC at $(\mathbf{r})$, respectively, with δC at $(\mathbf{r}')$, where the two fluctuations are separated by a time t:

$$\begin{aligned} \mathcal{A}_c(\mathbf{r}, \mathbf{r}', z, t) &\equiv \left\langle \delta A(\mathbf{r}, z, t) \delta C(\mathbf{r}', 0) \right\rangle \\ \mathcal{C}_c(\mathbf{r}, \mathbf{r}', t) &\equiv \left\langle \delta C(\mathbf{r}, t) \delta C(\mathbf{r}', 0) \right\rangle. \end{aligned} \tag{6.29}$$

Combining equations (6.23), (6.28), and (6.29) yields

$$G_c(t) \equiv Q^2 I_0^2 \iint \zeta(\mathbf{r}) \zeta(\mathbf{r}') \mathcal{C}_c(\mathbf{r}, \mathbf{r}', t) d^2\mathbf{r} d^2\mathbf{r}'. \tag{6.30}$$

To solve for $G(t)$ in either the TIR/FRAP or TIR/FCS form, we begin with the differential equations governing $A(\mathbf{r}, z, t)$ and $C(\mathbf{r}, z, t)$:

$$\begin{aligned} \frac{\partial A}{\partial t} &= D_A \nabla^2_{\mathbf{r},z} A \\ \frac{\partial C}{\partial t} &= D_C \nabla^2_{\mathbf{r}} C + k_1 A_{z\to 0} B - k_2 C, \end{aligned} \tag{6.31}$$

where D_A and D_C are the bulk and surface diffusion coefficients of the solute molecules and $A_{z\to 0} \equiv \lim_{z\to 0} A(\mathbf{r}, z, t)$. Note that the rate of change of surface concentration depends on the local bulk concentration A only at the $z = 0$ surface. The operators $\nabla^2_{\mathbf{r}}$ and $\nabla^2_{\mathbf{r},z}$ are two- and three-dimensional Laplacians, respectively.

The number of adsorptions per area per time and desorptions per area per time are $k_1 A_{z\to 0} B$ and $k_2 C$, respectively. The net flux of molecules to the surface from the solution is the difference between these two terms, and is equal to $D_A(\partial A/\partial z)_{z\to 0}$(according to Fick's Law). Therefore, we obtain a boundary condition:

$$D_A \left(\frac{\partial A}{\partial z} \right)_{z\to 0} = k_1 A_{z\to 0} B - k_2 C. \tag{6.32}$$

To solve for both $G_p(t)$ and $G_c(t)$ concurrently, equations (6.31) and (6.32) must be transformed into equations describing the variables $\mathcal{A}_p$, $\mathcal{C}_p$ and $\mathcal{A}_c$, $\mathcal{C}_c$.

For **TIR/FRAP**, we substitute expressions for A and C in terms of $\mathcal{A}$, $\mathcal{C}$, $\overline{A}$, and $\overline{C}$ (by substituting equation (6.29) into equations (6.31) and (6.32)) and use the equilibrium relationship equation (6.21). Although the relative concentration of bleached vs unbleached fluorophore adsorbed to the surface changes with time during TIR/FPR recovery, the concentration of free binding sites remains constant at the equilibrium value $\overline{B}$ during fluorescence recovery. $B(\mathbf{r}, t)$ in equation (6.31) is therefore replaced by $\overline{B}$.

For **TIR/FCS**, we write A, B, and C in terms of $\overline{A}$, $\overline{B}$, $\overline{C}$, δA, δB, and δC and substitute these expressions into equations (6.31) and (6.32). The total concentration of surface sites (free sites B plus bound sites C) remains exactly constant during an experiment; thus $\delta B = -\delta C$. After applying the equilibrium relationship equation (6.21), and eliminating a term proportional to the product of two fluctuations, we multiply the resultant equations by $\delta C(\mathbf{r}',0)$ and take the thermodynamic ensemble average on both sides of each equation.

The resulting differential equations for TIR/FPR and TIR/FCS are formally identical, as follows (with subscripts p and c suppressed):

$$\begin{aligned} \frac{\partial \mathcal{A}}{\partial t} &= D_A \nabla^2_{\mathbf{r},z} \mathcal{A} \\ \frac{\partial \mathcal{C}}{\partial t} &= D_C \nabla^2_{\mathbf{r}} C + k_1 \overline{B} \mathcal{A}_{z\to 0} - R_R \mathcal{C} \\ D_A \left(\frac{\partial \mathcal{A}}{\partial z} \right)_{z\to 0} &= \overline{B} \mathcal{A}_{z\to 0} - R_R \mathcal{C}, \end{aligned} \tag{6.33}$$

where

$$R_R \equiv \begin{cases} k_2 & \text{for TIR/FRAP} \\ k_1 \overline{A} + k_2 & \text{for TIR/FCS} \end{cases}. \tag{6.34}$$

The subscript R (for reaction) denotes the dependence of the rate on the adsorption/desorption reaction kinetics. One of the goals of TIR/FRAP and TIR/FCS experiments is to measure R_R.

We now consider boundary and initial conditions.

In **TIR/FRAP**, only the surface bound molecules within a finite surface area are photobleached, so that $C(\mathbf{r}, t)$ and $A(\mathbf{r}, z, t)$ far away from the bleached region never depart from their equilibrium values during fluorescence recovery:

$$\begin{aligned} \left[\mathcal{C}_p(\mathbf{r}, t)\right]_{|\mathbf{r}|\to\infty} = \left[\mathcal{A}_p(\mathbf{r}, z, t)\right]_{|\mathbf{r}|\to\infty} &= 0 \\ \left[\mathcal{A}_p(\mathbf{r}, z, t)\right]_{z\to\infty} &= 0. \end{aligned} \tag{6.35}$$

Immediately after bleaching the surface bound fluorophore at time $t = 0$, the unbleached bulk concentration $[A(\mathbf{r}, z, t)]_{t\to 0} = \overline{A}$, so

$$\left[\mathcal{A}_p(\mathbf{r}, z, t)\right]_{t\to 0} = 0. \tag{6.36}$$

The concentration $[C(\mathbf{r}, t)]_{t\to 0}$ of unbleached fluorophore on the surface (for a short bleaching pulse) is assumed to be an exponentially-decaying function of local laser intensity during the bleaching pulse, so that:

$$\left[\mathcal{C}_p(\mathbf{r}, t)\right]_{t\to 0} = \overline{C}(1 - e^{-K\zeta(\mathbf{r})}), \tag{6.37}$$

where K is a product of the intrinsic bleaching efficiency of the fluorophore and the bleaching duration and power. Using equations (6.26) and (6.37), the initial value of $G_p(t)$ is

$$G_p(0) = \overline{C}QI_0 \int \zeta(\mathbf{r})(1 - e^{-K\zeta(\mathbf{r})})d^2\mathbf{r}. \tag{6.38}$$

In **TIR/FCS**, the correlation between concentration fluctuations at two positions infinitely separated in space must be equal to zero, so that

$$\begin{aligned} &[\mathcal{C}_c(\mathbf{r}, \mathbf{r}', t)]_{|\mathbf{r}-\mathbf{r}_i|\to\infty} = [\mathcal{A}_c(\mathbf{r}, \mathbf{r}', z, t)]_{|\mathbf{r}-\mathbf{r}_i|\to\infty} = 0 \\ &[\mathcal{A}_c(\mathbf{r}, \mathbf{r}', z, t)]_{z\to\infty} = 0. \end{aligned} \tag{6.39}$$

The concentration $A(\mathbf{r})$ is defined within a small subvolume at location $\mathbf{r}$. This subvolume is open to an infinite reservoir of molecules in the bulk solution. The subvolume thereby contains a variable number of molecules which follows a Poisson distribution. A concentration fluctuation δA from equilibrium in the subvolume is uncorrelated with a concentration fluctuation δC on the surface at the same time:

$$[\mathcal{A}_c(\mathbf{r}, \mathbf{r}', z, t)]_{t\to 0} = 0. \tag{6.40}$$

(Note, however, that a fluctuation in A can be correlated with a fluctuation in C at a different time.) For a finite number of surface binding sites, the average of the square of the number fluctuation of adsorbed solute molecules within the observation region is equal to the product of the mean number of adsorbed molecules in this area and the average fraction β of the binding sites which are free, so that

$$[\mathcal{C}_c(\mathbf{r}, \mathbf{r}', t)]_{t\to 0} = \beta\overline{C}\delta(\mathbf{r} - \mathbf{r}'), \tag{6.41}$$

where

$$\beta \equiv \frac{\overline{B}}{\overline{B} + \overline{C}}. \tag{6.42}$$

Using equations (6.30) and (6.41), the initial value of $G_C(t)$ is

$$G_c(0) \equiv \beta\overline{C}Q^2I_0^2 \int \zeta^2(\mathbf{r})d^2\mathbf{r}. \tag{6.43}$$

We are now set to derive a solution for $G(t)$ valid for both TIR/FRAP and TIR/FCS. Equation (6.33) can be solved by linear transformation theory. We apply a Fourier transform with respect to the surface position vector $\mathbf{r} \to \mathbf{q}$ and a Laplace transform with respect to the normal to the surface $z \to p$ and time $t \to \omega$. In the following, transformations which have been performed are indicated solely by the variables in parentheses after the symbol $\mathcal{A}$ or $\mathcal{C}$.

The transformed states of equation (6.37) together with the boundary and initial condition equations (6.35), (6.37), (6.40), and (6.41) are:

$$\frac{\omega}{D_A}\mathcal{A}(\mathbf{q}, p, \omega) = (p^2 - q^2)\mathcal{A}(\mathbf{q}, p, \omega) - p[\mathcal{A}(\mathbf{q}, p, \omega)]_{z\to 0} - \left[\frac{\partial\mathcal{A}(\mathbf{q}, p, \omega)}{\partial z}\right]_{z\to 0} \tag{6.44}$$

$$\omega\mathcal{C}(\mathbf{q}, \omega) - [\mathcal{C}(\mathbf{q}, t)]_{t\to 0} = -D_C q^2\mathcal{C}(\mathbf{q}, \omega) + k_1\overline{B}[\mathcal{A}(\mathbf{q}, z, \omega)]_{z\to 0} - R_R\mathcal{C}(\mathbf{q}, \omega) \tag{6.45}$$

$$D_A\left[\frac{\partial\mathcal{A}(\mathbf{q}, z, \omega)}{\partial z}\right]_{z\to 0} = k_1\overline{B}[\mathcal{A}(\mathbf{q}, z, \omega)]_{z\to 0} - R_R\mathcal{C}(\mathbf{q}, \omega). \tag{6.46}$$

Solving equation (6.44) for $\mathcal{A}(\mathbf{q}, z, \omega)$ and applying an inverse Laplace transform from p-space back to z space yields

$$\begin{aligned}&\mathcal{A}(\mathbf{q}, z, \omega) = \\ &\frac{1}{2}\left\{[\mathcal{A}(\mathbf{q}, z, \omega)]_{z\to 0} + \left(q^2 + \frac{\omega}{D_A}\right)^{-1/2}\left[\frac{\partial\mathcal{A}(\mathbf{q}, z, \omega)}{\partial z}\right]_{z\to 0}\right\}e^{\left(q^2+\frac{\omega}{D_A}\right)^{1/2}z} + \\ &\frac{1}{2}\left\{[\mathcal{A}(\mathbf{q}, z, \omega)]_{z\to 0} - \left(q^2 + \frac{\omega}{D_A}\right)^{-1/2}\left[\frac{\partial\mathcal{A}(\mathbf{q}, z, \omega)}{\partial z}\right]_{z\to 0}\right\}e^{-\left(q^2+\frac{\omega}{D_A}\right)^{1/2}z}.\end{aligned} \tag{6.47}$$

We choose $\mathrm{Re}(q^2 + \omega/D_A) \geqslant 0$ and apply the boundary conditions to $\mathcal{A}$ as $z \to \infty$ (equations 6.35 and 6.39), which then requires that

$$[\mathcal{A}(\mathbf{q}, z, \omega)]_{z\to 0} = -\left(q^2 + \frac{\omega}{D_A}\right)^{-1/2}\left[\frac{\partial\mathcal{A}(\mathbf{q}, z, \omega)}{\partial z}\right]_{z\to 0}. \tag{6.48}$$

Equation (6.48) together with equation (6.46) yield a solution for $[\mathcal{A}(\mathbf{q}, z, \omega)]_{z\to 0}$ in terms of $\mathcal{C}(\mathbf{q}, \omega)$. This is substituted into equation (6.45) to arrive at a solution for $\mathcal{C}(\mathbf{q}, \omega)$:

$$\mathcal{C}(\mathbf{q}, \omega) = N(q, \omega)[\mathcal{C}(\mathbf{q}, t)]_{t\to 0}, \tag{6.49}$$

where

$$N(q, \omega) = \frac{D_A\left(q^2 + \frac{\omega}{D_A}\right)^{1/2} + k_1\overline{B}}{(\omega + R_R + q^2D_C)D_A\left(q^2 + \frac{\omega}{D_A}\right)^{1/2} + (\omega + q^2D_C)k_1\overline{B}}. \tag{6.50}$$

To arrive at $G_p(t)$ and $G_c(t)$, we put equation (6.49) together with the initial condition equations (6.37) and (6.41) into equations (6.26) and (6.30), thereby obtaining

$$G_p(t) = QI_0\int\zeta(\mathbf{r})L^{-1}_{\omega\to t}F^{-1}_{\mathbf{q}\to\mathbf{r}}\{N(q, \omega)F_{\mathbf{r}'\to\mathbf{q}}[\overline{C}(1 - e^{-K\zeta(\mathbf{r})})]\}d^2\mathbf{r} \tag{6.51}$$

and

$$G_c(t) = Q^2 I_0{}^2 \int \zeta(\mathbf{r})\zeta(\mathbf{r}') L^{-1}_{\omega\to t} F^{-1}_{\mathbf{q}\to\mathbf{r}}\left\{N(q,\omega) F_{\mathbf{r}''\to\mathbf{q}}[\beta\overline{C}\delta(\mathbf{r}'-\mathbf{r}'')]\right\} d^2\mathbf{r}\, d^2\mathbf{r}', \tag{6.52}$$

where F refers to a Fourier transform and $F^{-1}$and L^{-1} refer to inverse Fourier and Laplace transforms. Apart from multiplicative constants, $G_p(t)$ is identical to $G_c(t)$ in the limit of shallow photobleaching, K <<; 1. Referring to the expressions for the initial values of $G_p(t)$ and $G_c(t)$ (equations (6.42) and (6.47)), and applying Parseval's theorem, we rewrite $G(t)$ as

$$G(t) = G(0)\frac{L^{-1}_{\omega\to t}\int |\tilde{\zeta}(\mathbf{q})|^2 N(q,\omega) d^2\mathbf{q}}{\int |\tilde{\zeta}(\mathbf{q})|^2 d^2\mathbf{q}}, \tag{6.53}$$

where $\tilde{\zeta}(\mathbf{q})$ is the Fourier transform of $\zeta(\mathbf{r})$. Equation (6.53) is true for TIR/FCS in general and for TIR/FRAP when $K << 1$.

6.8.2 Characteristic rates

We wish to recast $G(t)$ in terms of physically significant characteristic rate parameters whose relative sizes determine the shape of $G(t)$. The form of the characteristic rate parameters can be obtained by converting the arguments $\mathbf{q}$ and ω in the integrand of equation (6.53) to unitless variables ξ and ν, as follows. We rewrite $\zeta(\mathbf{r})$ quite generally with unitless argument as

$$\zeta(\mathbf{r}) = \frac{1}{s^2}\zeta'(\mathbf{r}/s) \tag{6.54}$$

where s is a characteristic linear dimension of the observed region and $\zeta'(\mathbf{r}/s)$ is an intensity profile function with unitless argument. Then, the Fourier transform of equation (6.54) is

$$\tilde{\zeta}(\mathbf{q}) = \tilde{\zeta}'(\boldsymbol{\xi}) \tag{6.55}$$

where $\boldsymbol{\xi} \equiv \mathbf{q}s$ and $\tilde{\zeta}'(\boldsymbol{\xi})$is the Fourier transform of $\zeta'(\mathbf{r}/s)$. In addition, we define $\nu \equiv \omega/R_R$. From equations (6.21), (6.50), (6.53), and (6.55) we obtain

$$G(t) = G(0)\frac{L^{-1}_{\nu\to R_R}\int |\tilde{\zeta}'(\boldsymbol{\xi})|^2 N(\boldsymbol{\xi},\nu) d^2\boldsymbol{\xi}}{\int |\tilde{\zeta}'(\boldsymbol{\xi})|^2 d^2\boldsymbol{\xi}} \tag{6.56}$$

where

$$N(\boldsymbol{\xi},\nu) = \frac{\left(\nu+\xi^2\frac{R_{BLD}}{R_R}\right)^{1/2} + \left(\frac{R_R}{R_{BND}}\right)^{1/2}}{\left(1+\nu+\xi^2\frac{R_{SD}}{R_R}\right)\left(\nu+\xi^2\frac{R_{BLD}}{R_R}\right)^{1/2} + \left(\nu+\xi^2\frac{R_{SD}}{R_R}\right)\left(\frac{R_R}{R_{BND}}\right)^{1/2}} \tag{6.57}$$

and

$$R_R \equiv \begin{cases} k_2 & \text{for TIR/FRAP} \\ k_1\bar{A} + k_2 & \text{for TIR/FCS} \end{cases} \qquad \textbf{reaction rate} \qquad (6.58)$$

$$R_{BND} \equiv \begin{cases} \dfrac{D_A}{(\bar{C}/\bar{A})^2} & \text{for TIR/FRAP} \\ \dfrac{D_A}{(\beta\bar{C}/\bar{A})^2} & \text{for TIR/FCS} \end{cases} \qquad \textbf{bulk normal diffusion rate} \qquad (6.59)$$

$$R_{BLD} \equiv \frac{D_A}{s^2} \qquad \textbf{bulk lateral diffusion rate} \qquad (6.60)$$

$$R_{SD} \equiv \frac{D_C}{s^2} \qquad \textbf{surface diffusion rate} \qquad (6.61)$$

These four parameters are characteristic rates which describe distinct physical processes.

Because of the complexity of equations (6.56) and (6.57), we present here a qualitative discussion of the expected behavior of $G(t)$ in terms of its characteristic rates, before resuming a more formal development in the next section. The qualitative discussion refers to figure 6.19.

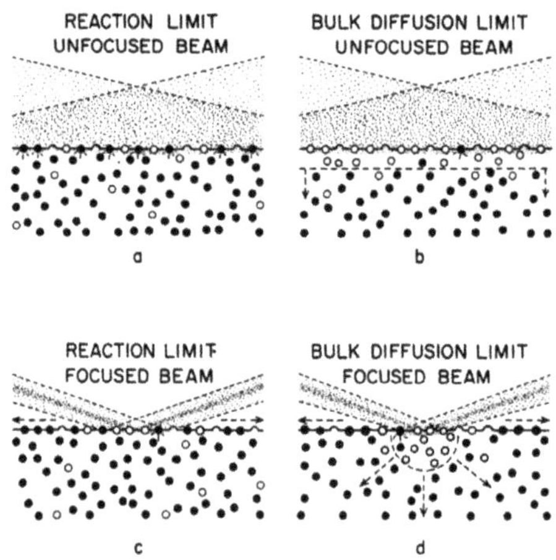

Figure 6.19. Conceptual illustration of surface kinetics. The reaction limit vs. bulk diffusion limit is shown with particular reference to TIR/FRAP. The concentrations of bleached (open circles) vs. unbleached (filled circles) as a function of position are schematically depicted at a short time after photobleaching. We assume the same kinetic constants k_1 and k_2 and the same time t after photobleaching in each panel (a-d); only the bulk diffusion constant D_A and the observation size s are varied. (a) Reaction limit, $s \to \infty$. The bleached solute has diffused far from the surface and has become uniformly diluted in the bulk solution. (b) Bulk diffusion limit, $s \to \infty$. The bleached solute remains in a 'cloud' near the surface. The slow diffusive dispersal of the cloud (shown bordered by a dotted line) limits the rate of fluorescence recovery. (c) Reaction limit, s finite. The bleached solute rapidly becomes diluted uniformly in the bulk solution, as in (a); surface diffusion provides a parallel route for fluorescence recovery. (d) Bulk diffusion limit, s finite. The diffusive dispersal of the cloud of bleached solute in the bulk limits the fluorescence recovery rate, as in (b); however, the cloud can disperse more rapidly due to the three-dimensional mode of escape. Surface diffusion speeds the recovery and alters the shape of the bleached cloud. Reprinted from Thompson *et al* (1981), copyright (1981), with permission from Elsevier.

6.8.3 R_R

To explain the significance of R_R, we consider a very large observation area $s \to \infty$, so that diffusion rates parallel to the surface (R_{BND}, R_{SD}) are negligibly small. We also assume that diffusion in the bulk is very fast (characterized by a large D_A). In TIR/FRAP, as adsorbed bleached molecules desorb, they quickly diffuse away from the region near the surface (figure 6.19a). The probability of a desorbed bleached molecule readsorbing to the surface within the duration of the experiment is very small. The system is analogous to a unimolecular isomerization reaction between adsorbed (C) and desorbed (A) unbleached solute molecules, where the number of A molecules is very large and behaves as an infinite reservoir to C molecules. The relaxation rate of this reaction (and thus the rate of fluorescence recovery) is k_2, which is just R_R for TIR/FRAP. In TIR/FCS, a surface concentration fluctuation relaxes by adsorption/desorption, becoming a bulk concentration fluctuation which quickly diffuses away from the surface. Because the total number of binding sites ($B + C$) remains constant during the relaxation process, this system is analogous to a unimolecular isomerization reaction between bound (C) and unbound (B) sites on the surface. The relaxation rate of this reaction (and thus the average rate of fluctuation relaxation) is $k_1\overline{A} + k_2$, which is just R_R for TIR/FCS. For both TIR/FRAP and TIR/FCS, we expect G(t) to decay with the characteristic rate R_R. This case of fast bulk diffusion and relatively slow adsorption/desorption is called the 'reaction limit.'

6.8.4 R_{BND}

At the other extreme from the reaction limit, bulk diffusion is slow; this is called the 'bulk diffusion limit'

In this case, for TIR/FRAP, adsorbed bleached molecules diffuse only very slowly from the vicinity of the bleached surface after desorption, resulting in a long-lived 'cloud' of bleached solute immediately adjacent to the surface (figure 6.19b). Some of the bleached solute is readsorbed to the surface, thereby slowing fluorescence recovery. The rate of recovery is no longer governed by the desorption rate k_2, but instead by the bulk diffusion coefficient D_A. The characteristic rate is determined by how rapidly bleached molecules on the surface can be completely replaced by an equal number of unbleached molecules from the bulk. For a (large) surface area S, the number of molecules on the surface is CS; an equal number $\overline{A}\,Sl_p$ will be found within some characteristic distance l_p from the surface if $l_p = \overline{C}/\overline{A}$. The characteristic rate R_{BND} for TIR/FRAP then becomes D_A/l_p^2.

For TIR/FCS in the bulk diffusion limit, length l_c(with subscript 'c') replaces l_p. The length l_c is determined by how rapidly a surface concentration fluctuation in C relaxes through exchange with a bulk concentration fluctuation in A. Consider, as above, a volume defined by a given surface area extended into the solution by a distance l_c. Because this volume is open to the transport of solute molecules, the average number fluctuation of A at equilibrium is given by $\langle \delta N_A{}^2 \rangle = \langle \overline{N_A} \rangle = \overline{A}\,Sl_c$. Because the total number of binding sites is finite, the average number fluctuation of C at equilibrium is given by $\langle \delta N_C{}^2 \rangle = \beta \langle \overline{N_C} \rangle = \beta \overline{C} S$. The length l_c for which the average number fluctuation of solute molecules $\langle \delta N_A{}^2 \rangle$ is equal to the average number fluctuation of surface bound molecules $\langle \delta N_C{}^2 \rangle$ is calculated from $\beta \overline{C} S = \overline{A} S l_c$, which gives $l_c = \beta \overline{C} / \overline{A}$. The characteristic rate R_{BND} for TIR/FCS then becomes D_A/l_c^2. For both TIR/FRAP and TIR/FCS, R_{BND} can thus be interpreted as the bulk diffusion rate normal to the surface across a depth of solution l. When bulk diffusion across l is slow, R_{BND} determines the decay rate of $G(t)$. In the bulk diffusion limit, $G(t)$ becomes sensitive to the bulk solution properties (D_A and $\overline{A}$) rather than to the surface reaction rate R_R.

6.8.5 R_{SD}

R_{SD} is the rate of surface diffusion (with coefficient D_C) through an area with a finite characteristic length s. In general, surface diffusion speeds the decay of $G(t)$.

6.8.6 R_{BLD}

R_{BLD} is the rate by which already desorbed molecules can exit the vicinity of the surface observation area by lateral diffusion in the bulk (i.e. parallel rather than normal to the surface). An increasing R_{BLD} (due to $s \to 0$) allows for faster dissipation of the 'cloud' of bleached solute (for TIR/FPR, see figure 6.19(c) and (d)) or concentration fluctuation (for TIR/FCS) in the vicinity of the observed region. This effect tends to bring $G(t)$ farther from the bulk diffusion limit and closer to the reaction limit.

6.8.7 Limiting solutions for an infinite observation area

In this section, we examine the limit behavior of $G(t)$ for very large areas as the characteristic rate ratios are varied. For very large areas, $G(t)$ assumes a simple and useful explicit form. In this case, the diffusion rates parallel to the surface observation area (R_{BND}, R_{SD}) are negligibly small. The factors determining $G(t)$ are the reaction rate R_R and the net z direction diffusion rate R_{BND}. Upon rewriting for $s \to \infty$, equations (6.56) and (6.57) can be expressed as a sum of four partial fractions which can be readily inverse Laplace transformed to obtain the following real function:

$$G_{s\to\infty}(t) = \frac{G(0)}{\nu_-^{1/2} - \nu_+^{1/2}} \left[\nu_-^{1/2} w\left(-i\sqrt{\nu_+ R_R t}\right) - \nu_+^{1/2} w\left(-i\sqrt{\nu_- R_R t}\right) \right], \quad (6.62)$$

where

$$\nu_\pm^{1/2} \equiv \frac{1}{2}\left(\frac{R_R}{R_{BND}}\right)^{1/2}\left[-1 \pm \left(1 - 4\frac{R_{BND}}{R_R}\right)^{1/2}\right] \quad (6.63)$$

and

$$w(i\eta) \equiv e^{\eta^2}\mathrm{erfc}(\eta) \quad (6.64)$$

where the argument η can be complex. The 'w function' (sometimes called a 'Fabreeva function') is a standard form, the mathematical properties of which can be found in reference sources such as Abramowitz and Stegun (1972). Unlike equation (6.56), equation (6.62) is valid for all K values in TIR/FRAP. This solution for $G_{s\to\infty}(t)$ can also be directly obtained from differential equations containing no surface position dependence.

$G_{s\to\infty}(t)$ is graphed in figure 6.20 for several values of the ratio R_R/R_{BND}.

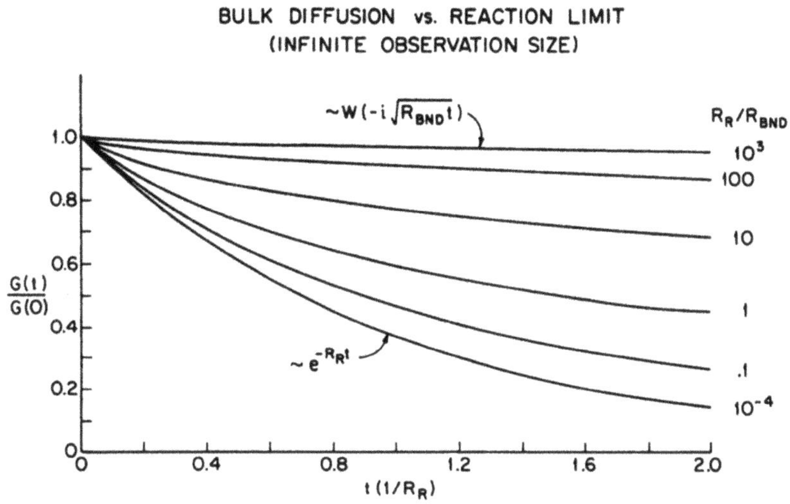

Figure 6.20. $G(t)$ for a very large observation area (s—) and variable ratio R_R/R_{BND} as R_R/R_{BND} approaches zero (the reaction limit). Time t is plotted in units of $1/R_R$. The normalized initial slope is always $-R_R$. Reprinted from Thompson *et al* (1981), copyright (1981), with permission from Elsevier.

The extremes of R_R/R_{BND} lead to simple analytic forms for $G_{s\to\infty}(t)$.

Reaction limit: if $R_R/R_{BND} << 1$, then $G_{s\to\infty}(t)$ depends only on R_R. This is the ideal state for measuring surface kinetic rates. In this case, $\nu_\pm^{1/2}$ approaches $\pm\, i$. Using a mathematical identity involving w functions, equation (6.62) approaches

$$G_{s\to\infty}(t) \to G(0)\exp(-R_R t). \tag{6.65}$$

Bulk diffusion limit: decreasing the bulk diffusion rate R_{BND} (for a fixed R_R) can only lead to a slower decay of $G_{s\to\infty}(t)$ than that given by equation (6.65). If $R_R/R_{BND} >> 1$, then $G_{s\to\infty}(t)$ depends only on R_{BND}. Under these conditions, the development of $G_{s\to\infty}(t)$ over time is related to bulk diffusion, not to surface kinetics, and the rates k_1 and k_2 cannot be easily obtained. (This problem can sometimes be overcome, as discussed in the next subsection.) In this case, equation (6.62) approaches

$$G_{s\to\infty}(t) \to G(0)w(-i\sqrt{R_{BND}t}). \tag{6.66}$$

6.8.8 Solutions for a finite observation area

In general, for a finite observation dimension s, $G(t)$ curves are computer-generated. However, in the special case of the reaction limit, a useful analytic function does exist. To simplify the presentation, we mainly discuss the low bleach limit ($K << 1$) in TIR/FRAP. In this limit, the shape of $G(t)$ is identical for both TIR/FRAP and TIR/FCS. However, the recovery curves for the deeper bleaching used in practical FRAP experiments (generally $K < 3$ for a Gaussian beam profile) deviate only slightly in shape and rate from those for $K << 1$.

Reaction limit ($R_R/R_{BND} \rightarrow 0$) *with finite surface area*: in this case, $G(t)$ depends only on R_R and the surface diffusion rate R_{SD}, regardless of the sizes of the illuminated and observed areas. Allowing $R_R/R_{BND} \rightarrow 0$ as R_{SD}/R_R remains finite in equations (6.56) and (6.57), then applying an inverse Laplace transform, Parseval's theorem, and the convolution theorem, $G(t)$ can be rewritten in terms of real space variables as:

$$\frac{G(t)}{G(0)} = e^{-R_R t} \frac{\iint \zeta(\mathbf{r})\zeta(\mathbf{r}') \exp\left(-\frac{|\mathbf{r}-\mathbf{r}'|^2}{4D_C t}\right) d^2\mathbf{r} d^2\mathbf{r}'}{4\pi D_C t \int \zeta^2(\mathbf{r}) d^2\mathbf{r}}. \tag{6.67}$$

Thus, $G(t)$ is the product of a simple exponential characteristic of the surface adsorption/desorption kinetics and a factor characteristic of surface diffusion through the observation area. Simple but experimentally practical shapes for $\zeta(\boldsymbol{r})$, such as elliptical, circular, or linear Gaussians, or squares, or sinusoidal periodic patterns, yield explicit forms of $G(t)$ upon straightforward integration. At the end of this subsection, an explicit solution is provided for a sinusoidal periodic pattern.

For deeper bleaching (i.e. higher K), replace $\zeta(\mathbf{r})$—but not $\zeta(\mathbf{r}')$—in equation (6.67) as follows: $\zeta(\mathbf{r}) \rightarrow [1 - e^{-K\zeta(\mathbf{r})}]/K$.

A more intuitively understandable approach can lead directly to equation (6.68). An FCS fluorescence autocorrelation function $G(\Delta\mathbf{r}, t)$ is basically the expectation value of the product of the fluorescence (expressed as a deviation from its equilibrium value) $\delta F(\mathbf{r}, t_1)$ at one location and time, with the fluorescence deviation $\delta F(\mathbf{r}', t_2)$ at a different location (where $\Delta\mathbf{r} = \mathbf{r} - \mathbf{r}'$) and time (where $t = t_2 - t_1$), presumably later:

$$G(\Delta\mathbf{r}, t) = \langle \delta F(\mathbf{r}, t_1)\delta F(\mathbf{r}', t_2)\rangle. \tag{6.68}$$

The expectation value is defined as the weighted average over all pairs of $(\mathbf{r}, \mathbf{r}')$ and (t_1, t_2). The weighting is determined by a 'joint probability density' $p(\mathbf{r}, \mathbf{r}', t)$, which is basically the probability that some molecule will be *both* at some location $\mathbf{r}$ at time t_1 and *also* (later) at location $\mathbf{r}'$ at time t_2. The weighted average can be written as an integral over both $\mathbf{r}$ and $\mathbf{r}'$, but restricted to locations for which $\Delta\mathbf{r} = \mathbf{r} - \mathbf{r}'$. Therefore, we can write (for the 2D geometry appropriate for TIRF):

$$G(\Delta\mathbf{r}, t) = \int \delta F(\mathbf{r}, t_1)\delta F(\mathbf{r}', t_2) p(\mathbf{r}, \mathbf{r}', t) d^2\mathbf{r} d^2\mathbf{r}'. \tag{6.69}$$

The joint probability density—the probability of being at 'point A' *and* some definite time later at 'point B'—is equal to the *conditional* probability of being at point B *given* that the molecule already had been at point A (at time t) earlier, multiplied by the probability of being at point A in the first place:

$$p(\mathbf{r}, \mathbf{r}', t) = p(\mathbf{r}'|\mathbf{r}, t)p(\mathbf{r}). \tag{6.70}$$

For systems in which the concentration of molecules is constant over space and time, $p(\mathbf{r})$ is a constant proportional to the equilibrium concentration and so can be taken out in front of the integral in equation (6.68). The fluorescence deviation factors δF are proportional to the local illumination intensities ζ ($\mathbf{r}$) and $\zeta(\mathbf{r}')$, which have no time dependence for time-steady illumination. Therefore, equation (6.69) becomes (with all constant factors combined into the factor H):

$$G(\Delta\mathbf{r},\, t) = H\int \zeta(\mathbf{r})\zeta(\mathbf{r}')p(\mathbf{r}'|\mathbf{r},\, t)d^2\mathbf{r}d^2\mathbf{r}'. \tag{6.71}$$

The conditional probability factor is the well-known function that gives the probability that a molecule at a starting location $\mathbf{r}$ will be found at r' at time t later as a result of 2D diffusion denoted by the diffusion coefficient D_C. It is

$$p(\mathbf{r}'|\mathbf{r},\, t) = \frac{1}{4\pi D_c t}\exp\left(-\frac{|\,\mathbf{r}-\mathbf{r}'|^2}{4D_C t}\right). \tag{6.72}$$

Installing equation (6.72) into equation (6.71) and providing the normalization as indicated by the denominator of equation (6.67) (which cancels out the K factor) leads directly to equation (6.67).

Bulk diffusion limit and intermediate cases: (($R_R/R_{BND}\not\to 0$)) *with finite surface area*: in these cases, equations (6.56) and (6.57) can be digitally integrated by computer. In general, for a fixed finite s, the reaction limit can be approached if $R_R/R_{BND} \to 0$. In terms of experimental variables, R_R/R_{BND} may be decreased by increasing the bulk concentration A; an experimental increase of bulk concentration A always brings $G(t)$ closer to the reaction limit. However, concentration A should not be increased so far that the number of bulk solute molecules in the evanescent field is comparable to the number adsorbed to the surface.

Given a $G(t)$ which is near the bulk diffusion limit for a very large observation area, progressively reducing s also causes $G(t)$ to approach the reaction limit. However, the size s cannot be decreased to less than the optical resolution limit of the fluorescence detection system. A $G(t)$ which is already at the reaction limit for a large observation area remains at the reaction limit as s is reduced.

6.8.9 TIR/FCS/FRAP to measure the diffusion coefficient in solution

The above subsections always assume that the number of fluorescent molecules in the bulk (but still present in the evanescent field) is negligible compared to the number bound (perhaps reversibly) on the surface. At the complete opposite extreme, in which there is no surface binding at all, TIR/FCS/FRAP can measure the bulk diffusion rate D_A of solute molecules near, but not actually contacting, the surface. This is interesting from a physical chemistry point of view, because the proximity of a surface can alter the local concentration of solutes and the viscosity and pH in the nearby solution.

We assume that the concentration of fluorescent molecules in the bulk solution (in the evanescent field near the surface) is A, which can fluctuate around the equilibrium value of $\overline{A}$. The average number of molecules N_A in an $h \times h$ square lateral area under observation is

$$N_A = \bar{A}h^2 \int_0^\infty e^{-z/d}dz = \bar{A}h^2 d. \tag{6.73}$$

We define the rate at which molecules diffuse through the evanescent field as $R_e = D_A/d^2$. The autocorrelation function $G(t)$ can be derived using the methods described in Starr and Thompson (2001):

$$G(t) = \frac{1}{2N_A}[(1-2R_e t)w(i\sqrt{R_e t}) + 2\sqrt{R_e t/\pi}], \tag{6.74}$$

where the w function is as defined in equation (6.64). (Equation (6.74) is a special case of the more general solution for combined surface adsorption/desorption kinetics and bulk diffusion in the evanescent field, as derived by Starr and Thompson.) Equation (6.74) is only appropriate if $h \gg d$ (which is generally the case), so that the probability of first escape by a molecule across the perimeter of the $h \times h$ square in the x–y plane is quite negligible compared to the probability of its escape through the evanescent field's z direction. Nevertheless, h should not be made too large, because N_A then becomes large and the amplitude of $G(t)$ might decrease to less than that of the systemic noise (light source fluctuations, etc.) in the measurement system.

The $G(t)$ expression (equation 6.74) looks a bit strange, with its last term on the right exploding as $t \to \infty$. However, the w function approaches $(\pi R_e t)^{-1/2}$ as $t \to \infty$, so that last term is canceled by the middle term, and then $G(t) \to (\pi R_e t)^{-1/2}$ as $t \to \infty$. This is a much slower decay than exponential decay. Because of the unfamiliarity of the w function, $G(t)$ is graphed in figure 6.21.

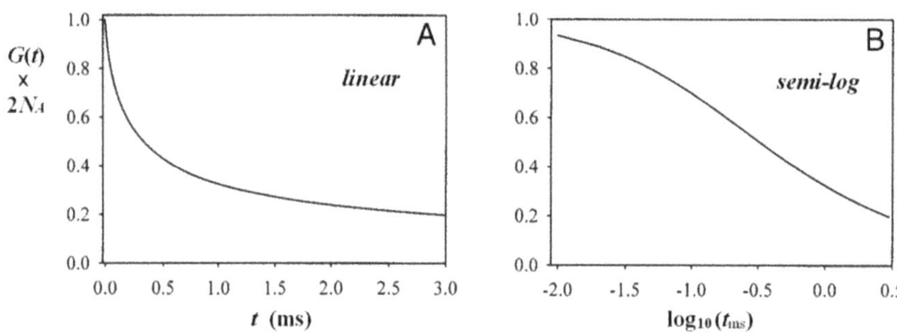

Figure 6.21. $G(t)$ in the limit of no surface binding, with R_e set at 1.02×10^4 s^{-1}. Panel A is a linear plot; panel B is a semi-log plot, in which the abscissa is the common log of the time expressed as a multiple of milliseconds. Reprinted from Starr *et al* (2001), copyright (2001), with permission from Elsevier.

The ratio of the initial slope to the initial value is R_e, which thereby provides a quick way of measuring D_C. $G(t)$ decreases to half of its initial value at 3.3 R_e, thereby providing an alternate way of measuring D_C.

6.8.10 Absolute concentrations: single component

FCS (not just in its TIR form) has a particular feature that FRAP does not have: the ability to report absolute concentrations. In any well-defined open volume (i.e. with an infinite reservoir outside the volume) that contains, on average, $\overline{N} \equiv \langle N \rangle$ molecules, random transport in and out of the volume creates Poisson fluctuations in N with a variance of

$$\mathrm{var}N \equiv \langle(\delta N)^2\rangle \equiv \langle(N - \overline{N})^2\rangle = \overline{N}. \tag{6.75}$$

In other words, the variance of the random fluctuations is equal to the average number of molecules. The fluorescence signal, however, only reports a value that is proportional to a generalized $N(t)$, written here as a 3D integral over a 3D illumination intensity profile $\zeta_{3D}(\mathbf{r})$:

$$F(t) = QI_0 \int \zeta_{3D}(\mathbf{r})\ C(\mathbf{r}, t)d^3\mathbf{r} = QI_0N(t), \tag{6.76}$$

where (see equation 6.23) Q is a product of the efficiencies of excitation light absorption and fluorescence emission and detection. The mean of F is:

$$\overline{F} = QI_0\overline{N}. \tag{6.77}$$

From equation (6.27), the autocorrelation function of F can be rewritten as

$$G_c(t) \equiv \left\langle \delta F(t)\delta F(0)\right\rangle = Q^2I_0^2\left\langle \delta N(t)\delta N(0)\right\rangle. \tag{6.78}$$

The $t \to 0$ value of $G_c(t)$ then becomes (apart from the contribution of photon shot noise at exactly $t = 0$)

$$G_c(t \to 0) \equiv Q^2I_0^2\left\langle \delta N(0)^2\right\rangle = Q^2I_0^2\overline{N}. \tag{6.79}$$

Equation (6.79) states that $G_c(t \to 0)$ is proportional to the variance of the fluctuations. It is highly significant that the extrapolation of $G_c(t)$ toward $t = 0$ is proportional to $\overline{N}$ and not $\overline{N}^2$(as is $\overline{F}^2$). Therefore, we can divide equation (6.78) by $\overline{F}^2$ (from equation 6.77) and cancel out the constants Q and I_0 (whose values are unknown), while still retaining an expression containing $\overline{N}$:

$$g(0) \equiv \frac{G_c(t \to 0)}{\overline{F}^2} = \frac{1}{\overline{N}}. \tag{6.80}$$

Lowercase g refers to the autocorrelation function as 'normalized' to $\overline{F}^2$.

The question may arise: why bother with the autocorrelation function G as $t \to 0$, rather than just compute the variance of experimental data directly from $\langle \delta F(t)\delta F(t)\rangle = \langle(\delta F)^2\rangle$? The reason is that photon shot noise in the data produces a large contribution to the variance of the fluorescence in typical FCS experiments. Fortunately, shot noise fluctuations among sample time bins of data are completely uncorrelated with each other. An 'up-fluctuation' due to shot noise in one sample bin has absolutely no predictive power over whether any of the next (or previous) sample bins will contain 'up' or 'down' fluctuations. Therefore, in an autocorrelation function $G(t)$, the shot noise only accumulates as a large spike at $t = 0$. (Although shot noise does not accumulate anywhere in G except at $t = 0$, shot noise does add unbiased uncertainty—i.e. noise—to G at all its time points. This effect can only be overcome by running the G averaging in equation (6.78) over as long a time as possible.) Extrapolating $G(t)$ back from its values at $t > 0$, where shot noise does not contribute in a biased way, we obtain a value for variance that depends only on the number fluctuations in $N(t)$.

The above formalism is written for the simplest case of FCS, in which $\overline{N}$ is the absolute average number of molecules in a volume (either 2D or 3D) with well-defined sharp edges, such as an open 3D volume. However, in actual experiments, the volume has 'soft' edges, as delimited by a focused laser spot in x and y or a decaying field in z. In addition, in the case of TIR/FCS for surface adsorption/desorption and surface diffusion, the number of molecules $N(t)$ is not open-ended: rather, it is 'capped' by the number of available surface binding sites. For such cases, an expression analogous to equation (6.80) can be derived (see equations 6.41–43):

$$\frac{G_c(t \to 0)}{\overline{F}^2} = \frac{\beta}{\overline{C}} \frac{\int \zeta^2(\mathbf{r}) d^2\mathbf{r}}{\left[\int \zeta(\mathbf{r}) d^2\mathbf{r}\right]^2}. \tag{6.81}$$

We can therefore calculate $\overline{C}$, given experimental values for $G_c(t \to 0)/\overline{F}^2$(from TIR/FCS) and β (from a Langmuir isotherm), where β is the average fraction of the surface binding sites that are free (thereby accounting for the 'capping'). The right-hand side of equation (6.81) can be rewritten as $1/\overline{N}_e$, where $\overline{N}_e$ is an 'equivalent' average number of molecules that can used in place of $\overline{N}$ as it appears in equation (6.80).

In the case in which TIR/FCS measures diffusion through the evanescent field (section b above), $N(t)$ is not capped by some theoretical maximum at all. The x–y region may be cleanly delimited by (say) an image-plane diaphragm or a camera pixel to an $h \times h$ area. However, in the z direction, some brightly illuminated molecules closer to the TIR surface 'count' for more than others farther away. In this case, the analogous expression for equation (6.81) is:

$$\frac{G_c(t \to 0)}{\overline{F}^2} = \frac{1}{\overline{C}} \frac{\int \zeta^2(\mathbf{r}) d^3\mathbf{r}}{\left[\int \zeta(\mathbf{r}) d^3\mathbf{r}\right]^2} = \frac{1}{\overline{C}h^2} \frac{\int_0^\infty e^{-2z/d} dz}{\left[\int_0^\infty e^{-z/d}\right]^2} = \frac{1}{2\overline{C}h^2 d}, \tag{6.82}$$

so that $\overline{N}_e = 2\overline{C}h^2 d$. This result makes intuitive sense. The product $h^2 d$ is the 'volume' in the evanescent field, and the factor of two accounts for the fact that the relative size of number fluctuations ($1/\overline{N}_e$) is reduced because many of the molecules are only dimly illuminated.

6.8.11 Absolute concentration: mixed components

It is still possible to determine absolute concentrations, even if the fluorescence arises from a mix of several components (say, monomers mixed with dimers). (Such mixtures are relevant, for example, in certain diseases, such as Alzheimer's disease, in which natural monomer proteins clump together.) 'Higher-order' FCS analysis of the spontaneous fluorescence fluctuations can reveal the existence and concentrations of multiple components in a sample. In practice (because of noise considerations), this is most readily done with simple mixes, say with two components. The theory presented here is not specific to TIR/FCS but is applicable to FCS in general.

Let a sample consist of just two different fluorescent constituents which number N_1 and N_2 in a small FCS or TIR/FCS volume, each in equilibrium with an infinite reservoir. The fluorescence spectra of each are assumed to be similar, so spectroscopy cannot be used to distinguish between them. For simplicity, we will assume that the two species are just mixed physically; they do not directly interact with each other. The goal here is to use higher-order FCS to find N_1 and N_2. In this special case, number fluctuations in either constituent are completely uncorrelated with number fluctuations in the other. Each contributes to the total fluorescence F according to its own (unknown) product Q of quantum efficiencies and light collection efficiencies (and ignoring shot noise):

$$F(t) = Q_1 I_0 N_1(t) + Q_2 I_0 N_2(t) \tag{6.83}$$

$$\overline{F} = Q_1 I_0 \overline{N_1} + Q_2 I_0 \overline{N_2}, \tag{6.84}$$

where, as above, the bar over a variable means the average. The relative quantum efficiency can be expressed as (the unknown) α_2:

$$\alpha_2 \equiv \frac{Q_2}{Q_1}. \tag{6.85}$$

Following the approach in the previous subsection, in which only one constituent was involved,

$$g(0) \equiv \frac{G_c(t \to 0)}{\overline{F}^2} = \frac{\langle(\delta F)^2\rangle}{\overline{F}^2} \tag{6.86}$$

where

$$\begin{aligned} \overline{F} &= Q_1 I_0[\overline{N_1} + \alpha_2 \overline{N_2}] \\ \delta F &= Q_1 I_0(\delta N_1 + \alpha_2 \delta N_2) \\ \langle(\delta F)^2\rangle &= Q_1^2 I_0^2[\langle(\delta N_1)^2\rangle + \alpha_2^2\langle(\delta N_2)^2\rangle]. \end{aligned} \tag{6.87}$$

Cross terms involving $\langle \delta N_1 \delta N_1 \rangle$ are missing from the last equation of (6.87) because they are zero, under the assumption that fluctuations in $N_1(t)$ and $N_2(t)$ are independent. Each of $N_1(t)$ and $N_2(t)$ follows its own Poisson distribution, so that

$$\langle(\delta N_i)^2\rangle = \overline{N_i}; \tag{6.88}$$

therefore, equation (6.86) becomes:

$$g(0) = \frac{[\langle(\delta N_1)^2\rangle + \alpha_2^2\langle(\delta N_2)^2\rangle]}{(\overline{N_1} + \alpha_2 \overline{N_2})^2} = \frac{\overline{N_1} + \alpha_2^2 \overline{N_2}}{(\overline{N_1} + \alpha_2 \overline{N_2})^2}. \tag{6.89}$$

6.8.12 Higher order TIR-FCS

Presumably, $g(0)$ is known from an experiment, but there are three unknowns ($\overline{N_1}$, $\overline{N_2}$, and α_2) in equation (6.89). We must find two more independent equations to compute all those variables. The additional equations can be obtained from higher-order autocorrelation functions, which are based on the same observed data vs. time (so no additional experiments are required), but are computed according to:

$$g_{m,\,n}(t) = \frac{\left\langle \delta F^m(t_2)\delta F^n(t_1)\right\rangle}{\overline{F}^{m+n}}, \tag{6.90}$$

where $t = t_2 - t_1$. With this definition, the above variable $g(0)$ is simply $g_{1,1}(0)$. Here, we are interested in the extrapolation to $t = 0$ (so $t_2 = t_1$), thus only the sum $m + n$ matters:

$$g_{m+n}(0) = \frac{\left\langle \delta F^{m+n}\right\rangle}{\overline{F}^{m+n}}. \tag{6.91}$$

In this notation, the standard first-order autocorrelation variable calculated in equation (6.46) is just $g_2(0)$. To calculate the next higher-order autocorrelation functions g_3 and g_4 as $t \to 0$, we use the following facts about the moments of the Poisson distribution:

$$\begin{aligned} \langle(\delta N)^2\rangle &= \overline{N} \\ \langle(\delta N)^3\rangle &= \overline{N} \\ \langle(\delta N)^4\rangle &= \overline{N}(1+3\overline{N}) \end{aligned} \tag{6.92}$$

$$g_3(0) = \frac{\langle(\delta F)^3\rangle}{\overline{F}^3} = \frac{\overline{N_1} + \alpha_2{}^3\overline{N_2}}{(\overline{N_1} + \alpha_2\overline{N_2})^3} \tag{6.93}$$

$$g_4(0) = \frac{\langle(\Delta F)^4\rangle}{\langle F\rangle^4} = \frac{\overline{N_1}(1+3\overline{N_1}) + 6\alpha_2{}^2\overline{N_1}\,\overline{N_2} + \alpha_2{}^4\overline{N_2}(1+3\overline{N_2})}{(\overline{N_1} + \alpha_2\overline{N_2})^4}. \tag{6.94}$$

In deriving equations (6.93)–(6.94), the cross terms involving $\langle\delta N_1(\delta N_2)^k\rangle$or$\langle(\delta N_1)^k\delta N_2\rangle$ are zero, and $\langle(\delta N_1)^2(\delta N_2)^2\rangle = \langle(\delta N_1)^2\rangle\langle(\delta N_2)^2\rangle$ because fluctuations in $N_1(t)$ and $N_2(t)$ are assumed to be independent.

Equations (6.89), (6.93), and (6.94), for experimentally-determined values of $g_2(0)$, $g_3(0)$, and $g_4(0)$ are the three equations we need to unambiguously solve for $\overline{N_1}$, $\overline{N_2}$, and α_2. Qualitatively, this procedure works because a mixture of constituents causes the total fluorescence to depart from a pure Poisson distribution, and the higher-order autocorrelations allow the measurement of higher moments of the actual distribution.

6.8.13 TIR/FRAP in a sub-resolution confined volume: a spherical secretory granule

The theory in the previous section applied to situations in which the concentration A was in an open volume. However, in cell biology, finite volumes are just as likely. In this subsection, we present an application of TIR/FRAP for measuring the diffusion rates of proteins either on the surface of, or inside, a spherical secretory granule in a living cell.

6.8.13.1 Conceptual theory

A secretory granule is a hollow sphere with a lipid membrane shell. Its interior contains chemicals destined for secretion. Its membrane contains proteins presumably involved in fusion with the cell's plasma membrane as a necessary step toward secretion. The radius of a granule is significantly smaller (typically ~150 nm) than the wavelength of visible light (~400 nm). It is also smaller than a focused laser spot, so standard non-TIRF spot photobleaching to determine diffusion coefficients would not be fruitful. However, the TIRF evanescent field is typically 50–100 nm in depth, so illuminating—and photobleaching—just the substrate-proximal side of a granule is feasible.

Granules in living secretory cells, visualized by transfection with a membrane or luminal protein tagged with a GFP-type appendage, are photobleached using TIR excitation light. As depicted in figure 6.21, the gradient of bleached fluorophore imprinted by the brief (ideally 'instantaneous') bright evanescent flash dissipates over time if the protein is mobile, or remains stable if the protein is immobile. The redistribution is excited by the same evanescent field, but at a much reduced (by at least a factor of 100) 'probe' intensity. The rate of fluorescence recovery, monitored in a postbleach sequence of camera images, is proportional to the diffusion coefficient.

There are two versions of this experiment: one for fluorophore-labeled membrane proteins confined to 2D diffusion on the granule surface, and the other for fluorophore-labeled luminal proteins confined to 3D diffusion within the granule interior.

The qualitative interpretations and relevant complications of the two versions of this experiment are the same; the only difference lies in the quantitative theoretical interpretation. We can approach a theory for fluorescence recovery in one of two ways: an explicit integral form (outlined here only for the shell geometry shown on the left-hand side of figure 6.22) or a Monte Carlo simulation for either the shell or the bulk.

Explicit integral for TIR/FRAP time dependence on a spherical shell

Let $C_0 \equiv$ the concentration of (unbleached) fluorophore on the surface before bleaching.

Let $C(\Omega, t) \equiv$ the concentration of (unbleached) fluorophore on the surface after bleaching.

Let $\Delta C(\Omega, t) \equiv C_0 - C(\Omega, t) =$ the concentration of **bleached** fluorophore at angular position Ω on the spherical surface. The bleached fluorophore spreads diffusively on the spherical surface. $\Delta C(\Omega, t)$ can be written as an integral over the diffusive behavior of a single fluorophore starting at position Ω_0, multiplied by the actual initial concentration at $t = 0$. Mathematically, the diffusive spread of a δ function is a 'Green's function' $G(\Omega, \Omega_0, t)$.

$$\Delta C(\Omega, t) = \int \Delta C(\Omega_0, 0)\ G(\Omega, \Omega_0, t) d\Omega_0 \tag{6.95}$$

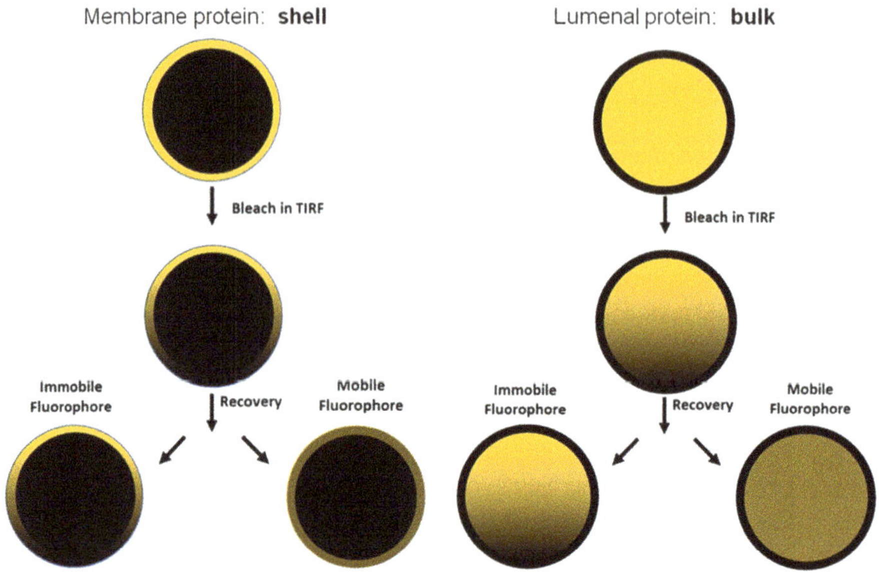

Figure 6.22. Schematic of the principle underlying the TIR/FRAP method used on a spherical secretory granule. The left-hand panel applies to membrane proteins (reprinted with permission from Abbineni *et al* (2021). The right-hand panel applies to proteins in the interior (the 'lumen') (reprinted from Ngatchou-Weiss et al (2014), copyright (2014), with permission from Elsevier. The TIR substrate (not shown) is tangent to the bottom of the sphere.

where

$$\Omega_0 = \left(\theta_0, \phi_0\right) \text{ and } d\Omega_0 = \sin\theta_0 d\theta_0 d\phi_0. \tag{6.96}$$

The initial postbleach concentration (for a bleach during T and $z = 0$ evanescent intensity I_0 is:

$$\Delta C(\Omega_0, 0) = C_0[1 - \exp(-\alpha T I_0 e^{-z_0/d})] = C_0[1 - \exp(-K e^{-z_0/d})], \text{ where } K \equiv \alpha T I_0 \tag{6.97}$$

and z_0 is the initial z position of the δ function, and the Green's function is

$$G(\Omega, \Omega_0, t) = \frac{1}{2}\sum_{n=0}^{\infty}(2n + 1)P_n(\cos\Delta\theta)e^{-n(n+1)Dt} \tag{6.98}$$

where

$$\Delta\theta \equiv \theta - \theta_0. \tag{6.99}$$

A side note: the Green's function shown in equation (6.8) is generally useful: it also describes the rotational diffusion of a molecule. Of particular interest here is an absorption or emission dipole moment vector 'attached' to the molecule. The 'tip of the arrow' of a rotationally diffusing constant-amplitude vector follows exactly the same mathematical rules as a point diffusing on a spherical surface.

The Legendre polynomial P_n can be written in terms of the associated Legendre polynomials P_n^m:

$$P_n(\cos\Delta\theta) = P_n(\cos\theta)P_n(\cos\theta_0) + 2\sum_{m=1}^{n}\frac{(n-m)!}{(n+m)!}P_n^m(\cos\theta)P_n^m(\cos\theta_0) \cos(m\Delta\phi), \tag{6.100}$$

where

$$\begin{aligned} \Delta\phi &\equiv \phi - \phi_0, \text{ so} \\ \cos(m\Delta\phi) &= \cos(m\phi)\cos(m\phi_0) + \sin(m\phi)\sin(m\phi_0). \end{aligned} \tag{6.101}$$

Note that both $\Delta C(\Omega_0, 0)$ (equation 6.97) and $G(\Omega, \Omega_0, t)$ (equation 6.98) have no dependence on ϕ_0. Therefore, the integration of ϕ_0 called for by equation (6.95) will be 'zeroed-out' by equation (6.101), except for the $m = 0$ term (for which it introduces the factor 2π). $\Delta C(\Omega, t)$ then becomes

$$\Delta C(\Omega, t) = \pi C_0 \int_{\theta_0=0}^{\pi}\left[1 - \exp\left(-Ke^{-z_0/d}\right)\right]\sum_{n=0}^{\infty}(2n+1)e^{-n(n+1)Dt}P_n(\cos\theta) P_n(\cos\theta_0)\sin\theta_0 d\theta_0. \tag{6.102}$$

For a granule sphere of radius R,

$$\begin{aligned} z &= R(1 - \cos\theta) \\ z_0 &= R(1 - \cos\theta_0). \end{aligned} \tag{6.103}$$

We also use the substitutions

$$\begin{aligned} x &\equiv \cos\theta \\ x_0 &\equiv \cos\theta_0 \\ B_n(t) &\equiv (2n+1)e^{-n(n+1)Dt} \\ f(x_0) &\equiv 1 - \exp\left(-Ke^{-(1-x_0)R/d}\right). \end{aligned} \tag{6.104}$$

Equation (6.102) becomes

$$\Delta C(\Omega, t) = \Delta C(x, t) = \pi\sum_{n=0}^{\infty}B_n(t)P_n(x)\int_{-1}^{1}f(x_0)P_n(x_0)dx_0. \tag{6.105}$$

The fluorescence on a spherical shell excited by a TIR probe of intensity $I_{\mathrm{TIR}}e^{-z/d}$ at any time is $F_{\mathrm{TIR}}^{\mathrm{shell}}(t)$

$$\begin{aligned} F_{\mathrm{TIR}}^{\mathrm{shell}}(t) &\equiv qI_{\mathrm{TIR}}\int e^{-z/d}C(\Omega, t)d\Omega \\ &= 2\pi qI_{\mathrm{TIR}}\int_{-1}^{1}e^{-(1-x)R/d}[C_0 - \Delta C(x, t)]dx. \end{aligned} \tag{6.106}$$

where q is a constant factor that includes all the efficiencies of emission and collection as well as unit conversions.

6.8.13.2 Monte Carlo simulation for TIR/FRAP on or in a sphere

The form of the recovery, as presented above, is complicated enough to require computer integration (involving sums over Legendre polynomials), and yet it is still oversimplified: there are several significant real-world factors which make the solution, and the ultimate determination of the diffusion coefficient, even more complicated:

- *Scattering.* The excitation itself is not a pure TIR-produced exponential decay, but some fraction of its intensity arises from scattering, originating either from the microscope optics or from refractive index irregularities in the sample itself. This scattering fraction can lead to its own bleaching during all phases of the experiment. The actual excitation intensity is therefore the sum of the exponentially-decaying evanescent part plus a z-independent scattering part, estimated to be 10% to 20% of the $z = 0$ evanescent intensity (see section 3.5).
- *Reversible bleaching.* The timescales of the experiments can be short, on the order of tens to hundreds of milliseconds. On those timescales, bleaching is not entirely irreversible; some bleached fluorophores spontaneously return to a ground state (e.g. return from a long-lived triplet state) but are capable of re-excitation to a fluorescence-producing excited state. This reversible recovery must be distinguished from diffusion-based recovery.

  The mechanism of reversible bleaching is not well known. Possible modes include the following. (1) Reversible and irreversible bleaching are parallel processes, either of which can occur in an unbleached fluorophore with some fixed likelihood. (2) A reversibly bleached fluorophore is protected against subsequent irreversible bleaching until it returns to the unbleached ground state. (3) Irreversible bleaching can occur only in a fluorophore which is already in a reversibly bleached state. Each of these possibilities would lead to a different dependence of the ratio of reversible to irreversible bleach depth as the incident illumination intensity is changed. The question of which mode is correct can be investigated by examining the reversible vs irreversible bleach depth under EPI illumination. (In EPI, the excitation intensity through the entire depth of the granule is constant, and diffusion-related effects thereby become irrelevant.) Observations have been most consistent with alternative (3): that irreversible bleaching occurs mainly to fluorophores while they are in a reversible bleached state. If experiments had taken place over a much longer timescale, reversible bleaching would not have been evident, but that would have been too long to have seen the diffusion effects that motivated the experiments (Abbineni *et al* 2022).
- *Bleaching during probe.* During the prebleach and postbleach dim 'probe' phase, significant bleaching (either reversible, irreversible, or both) can occur, which tends to cause a loss of fluorescence. Because the illumination is z dependent, even probe-phase bleaching occurs non-uniformly around the sphere's surface or in its interior.
- *Diffusion during bleach.* During the finite time duration of the bright 'bleach' phase, significant diffusion of the fluorophore can occur, making the $t = 0$ distribution of fluorophores immediately at the end of the bleach phase somewhat poorly defined.
- *Range of granule z positions.* Each granule may be located at a slightly different distance from the TIR/coverslip surface and thereby receives a different evanescent illumination intensity. This variability manifests itself as a different $t = 0$ postbleach fluorescence. This variability affects the postbleach distribution

of unbleached fluorophores and ultimately the exact form and timescale of the fluorescent recovery. For this reason, results should be grouped and averaged into several separate ranges of bleach depth.

To account for all these complications, a custom-written Monte Carlo-type program that repeatedly simulates the diffusion as a virtual random walk of a single fluorophore molecule can be used as a theoretical framework. A simulated membrane protein diffusing on a spherical surface, or within the sphere, is first positioned at a random location in its realm. During its walk, the fluorescence it produces over time is calculated and recorded from its z position, given a z-dependent excitation illumination intensity that includes both evanescent decay and scattering. This molecule is tracked over its entire random walk over time, during which it may be temporarily (reversibly) bleached or remain viable, until it irreversibly bleaches. Another molecule is then launched from a new random location. Altogether, a complete simulation consists of summing the fluorescence over time of tens of thousands of such single-molecule tracks.

6.8.13.3 Prebleach and immediate postbleach fluorescence: shell and bulk

A key question is whether the initial nonuniform distribution of unbleached fluorophores, as schematically depicted in figure 6.21, occurs at all for any particular granule. It is entirely possible that rapid granule tumbling, TIR scattering, or rapid protein diffusion may obviate the initial nonuniform distribution, and that the entire subsequent fluorescence recovery (if it happens at all) is due only to reversible bleaching. To resolve this question, the apparent bleach depth (induced by a TIR bleach pulse), recorded at the time of the first postbleach 'probe' observation, can be viewed in both EPI and in TIR (in practice, in rapid succession) by changing the incidence angle. In principle, for a granule with immobile fluorophore, a TIR probe interrogates the proximal part of the granule that has just been TIR bleached, so it should appear as a relatively deep bleach. An EPI probe interrogates all of the granule, including the more distal unbleached part, so it should appear as a relatively shallow bleach. On the other hand, a granule with highly mobile fluorophore redistributes its unbleached fluorophore well before the first probe observation, so that the TIR and EPI probe bleach depths appear the same. Figure 6.23 shows how the apparent bleach depth, as measured on a hypothetical recovery curve, depends upon fluorophore mobility.

For zero mobility (for which the difference between TIR and EPI bleach depth is the greatest), the expected prebleach and immediate postbleach fluorescence intensities can be written as simple integrals. The spherical shell fluorescence excited by a TIR probe before bleaching is $F_{\rm TIR}^{\rm shell}(-)$:

$$F_{\rm TIR}^{\rm shell}(-) = 2\pi q I_{\rm TIR} C_0 \int_{-1}^{1} e^{-(1-x)R/d} dx = \frac{2\pi q I_{\rm TIR} C_0 d}{R}(1 - e^{-2R/d}) \mid \qquad (6.107)$$

The shell fluorescence excited by a TIR probe of intensity $I_{\rm TIR}e^{-z/d}$ immediately after a very brief bleach pulse can be found by plugging $t = 0$ into equation (6.105). But it

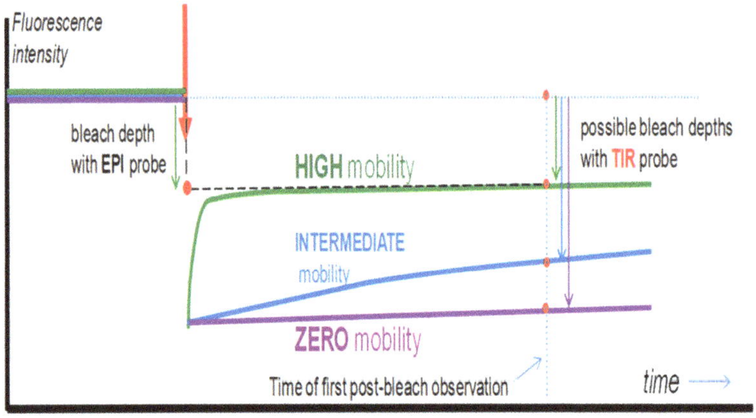

Figure 6.23. Schematic view of hypothetical TIR/FRAP recovery curves. Note that recovery never achieves 100% of the prebleach value because the volume (or surface area) of a small spherical granule is finite. The time of first postbleach observation may occur after some or all of the recovery has already occurred. The bleach pulse always uses TIR. The 'high' mobility (green) curve shows no difference between apparent bleach depths with EPI vs TIR probes. The 'zero' mobility case (purple) shows a maximal difference.

is more easily found by straightforward integration, without dealing with Green's functions, as follows:

$$F_{\text{TIR}}^{\text{shell}}(0) = qI_{\text{TIR}} \int_{z=0}^{2R} e^{-z/d} C(z)[2\pi R \sin\theta d\theta], \tag{6.108}$$

where the factor in the square brackets is the differential area of a ring on the surface of the sphere at position z, and where $C(z)$ is the immediate postbleach concentration of unbleached fluorophore on the spherical surface. From equations (6.102) and (6.96), $dz = R\sin\theta d\theta$ and $C(z) = C_0 \exp(-Ke^{-z/d})$, so

$$F_{\text{TIR}}^{\text{shell}}(0) = 2\pi qI_{\text{TIR}}C_0 \int_{z=0}^{2R} \exp\left[-\left(\frac{z}{d} + Ke^{-z/d}\right)\right]dz. \tag{6.109}$$

The shell fluorescence excited by an EPI probe of z-independent intensity I_{EPI} is similar to the above equations (6.106–6.109) for F_{TIR}, but without the factor $\exp(-z/d)$:

$$F_{\text{EPI}}^{\text{shell}}(t) \equiv qI_{\text{EPI}} \int C(\Omega, t)d\Omega = 2\pi qI_{\text{EPI}} \int_{-1}^{1} [C_0 - \Delta C(x, t)]dx \tag{6.110}$$

$$F_{\text{EPI}}^{\text{shell}}(-) = 2\pi qI_{\text{EPI}}C_0 \int_{-1}^{1} dx = 4\pi qI_{\text{EPI}}C_0 \tag{6.111}$$

$$F_{\text{EPI}}^{\text{shell}}(0) = 2\pi qI_{\text{EPI}}C_0 \int_{z=0}^{2R} \exp(-Ke^{-z/d})dz. \tag{6.112}$$

For TIR/FRAP on spheres filled with fluorophore in the interior ('bulk'), a complete integral form for the time dependence is not presented here. In that situation, the Green's function is more complicated, because its form depends on the radial position of the delta function. But as with the shell case, the pre- and immediate $t = 0$ postbleach fluorescence can be derived without it and instead integrated directly from the following:

$$F_{\mathrm{TIR}}^{\mathrm{bulk}}(-) = \pi q I_{\mathrm{TIR}} C_0 \int_{z=0}^{2R} e^{-z/d}(2zR - z^2)dz \tag{6.113}$$

$$F_{\mathrm{TIR}}^{\mathrm{bulk}}(0) = \pi q I_{\mathrm{TIR}} C_0 \int_{z=0}^{2R} e^{-z/d} \exp(-Ke^{-z/d})(2zR - z^2)dz \tag{6.114}$$

$$F_{\mathrm{EPI}}^{\mathrm{bulk}}(-) = \pi q I_{\mathrm{TIR}} C_0 \int_{z=0}^{2R} (2zR - z^2)dz \tag{6.115}$$

$$F_{\mathrm{EPI}}^{\mathrm{bulk}}(0) = \pi q I_{\mathrm{TIR}} C_0 \int_{z=0}^{2R} \exp(-Ke^{-z/d})(2zR - z^2)dz \tag{6.116}$$

The actual bleach depths as seen by TIR vs EPI probe are different monotonically increasing functions of TIR bleach-pulse intensity, and the bleach depth can vary from experiment to experiment or from granule to granule, depending on its z position. The theoretical bleach depths (as measured at the first postbleach observation time, which may be somewhat later than the end of the actual bleach) as probed by TIR can be plotted as a function of the EPI probe (figure 6.24). Experimental data from a number of granules can be plotted on these axes to see

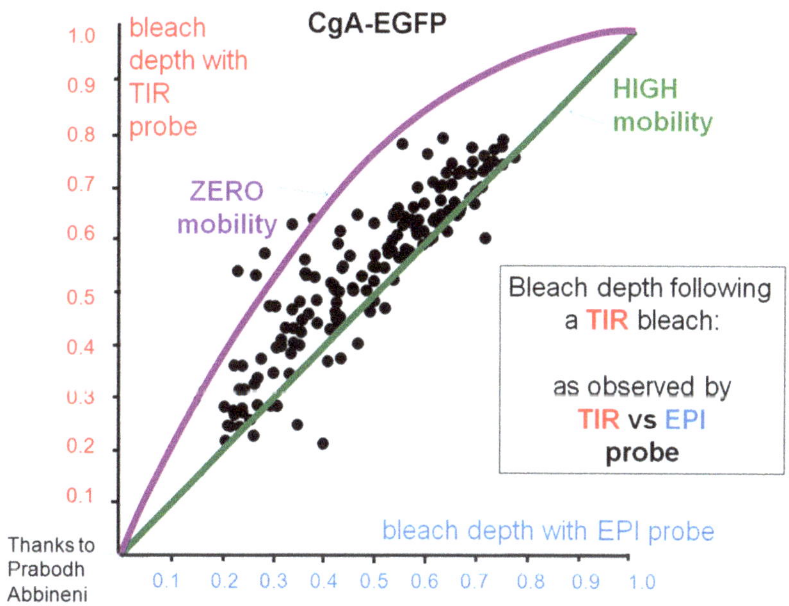

Figure 6.24. Bleaching depth for 'bulk' mobility as seen by TIR probe vs EPI probe for three cases: theoretical low-mobility fluorophore (purple); theoretical high-mobility fluorophore (green); and actual data for secretory granules (black), in this case, with its chromogranin A contents marked by EGFP. The fact that most of the experimental data points show a higher TIR apparent bleach depth than EPI apparent bleach depth shows that (a) the TIR bleach successfully imprinted a nonuniform pattern of bleaching, as schematically depicted in figure 6.21 and (b) that the mobility was slow enough to retain that pattern until at least the time of the first postbleach observation.

whether the TIR apparent bleach depth does indeed exceed the EPI apparent bleach depth. If it does, this is solid proof that the evanescent field did indeed imprint a nonuniform pattern on the granule. If the bleach depths are equal, then no strong conclusion can be drawn: either nonuniform imprinting did not happen, or the fluorophore mobility was high.

6.8.14 Spatially-resolved TIR/FRAP

TIR/FRAP, used in a spatially-resolved mode to measure binding/unbinding rates at the cell surface, only requires that the fluorescence recovery after a wide-area bleach be observed over time by a digital camera. The experimental recovery curves for each pixel can be fitted to the appropriate theoretical curve containing adjustable kinetic fit parameters. For example, in the reaction limit, the TIR/FRAP recovery rate provides a measure of the 'off-rate' $k_{\rm off}$ (see equations 6.38 and 6.69, where $k_2 = k_{\rm off}$). To measure $k_{\rm on}$, we only need to combine the TIR/FRAP recovery with a prior concentration jump relaxation of the same cell (see figure 6.25).

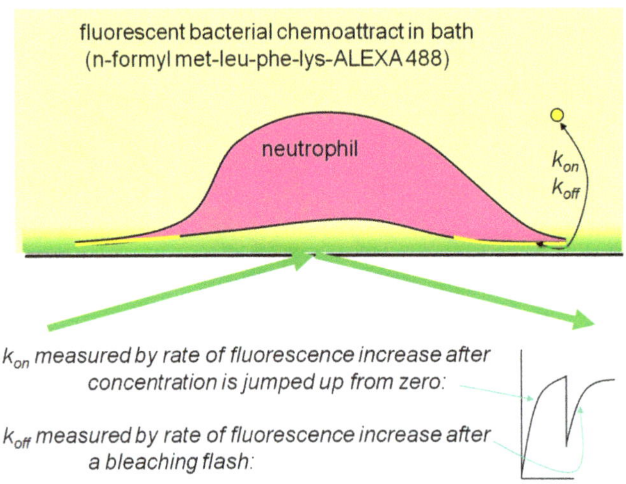

Figure 6.25. Schematic view of measuring the on/off kinetic rates at a cell surface near a cell edge.

Figure 6.26 shows typical experimental results, averaged over a few pixels, on the neutrophil. The FRAP part of the experiment is performed once the neutrophil reaches equilibrium with the fluorescence-marked chemoattractant.

After pixel-by-pixel curve-fitting analysis, a pseudo-image of the kinetic rates can be constructed. Figure 6.27 shows the result for $k_{\rm off}$. The results vary from place to place by several orders of magnitude. For more details see Axelrod and Omann (2006).

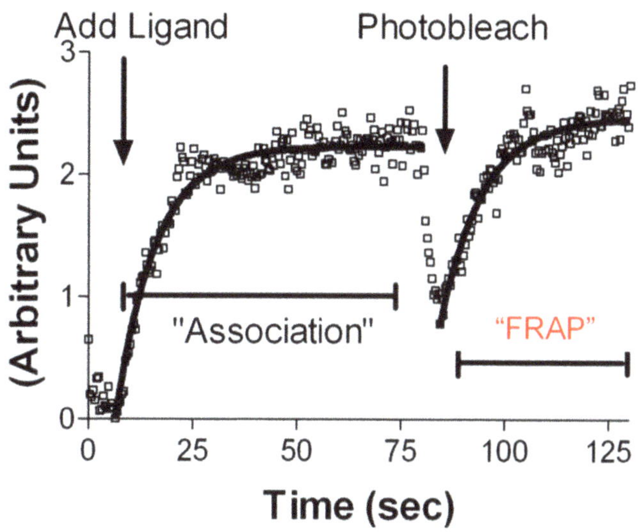

Figure 6.26. Sequential concentration jump and TIR/FRAP; actual data for the biological system identified in figure 6.24.

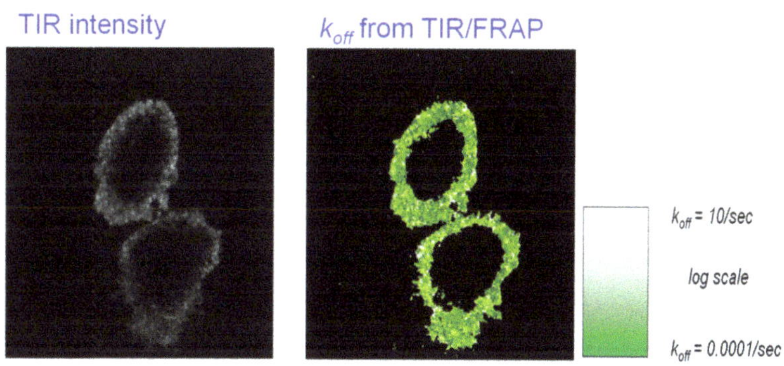

Figure 6.27. Pseudomap of k_{off}, based on pixel-by-pixel analysis. The diameter of a cell is ~15 μm. Black spots and regions in the k_{off} pseudomap are where the data was too dim and noisy for a good fit.

6.8.15 TIR/FRAP with sine wave

If a TIR periodic pattern (as discussed in section 5.11.3) is combined with FRAP, then antinode regions experience a much deeper bleach than node regions, setting up a periodic gradient of unbleached fluorophore. As this relaxes by diffusion, the same pattern (but much reduced in intensity) then probes fluorescence recovery. If the fluorescent material is only reversibly bound to the surface, the recovery is due to both adsorption/desorption kinetics and surface diffusion. This case is particularly interesting from a biological viewpoint, because surface diffusion can enhance the rate at which a nonspecifically and reversibly adsorbed solute molecule might find its specific receptor on a cell surface (Adam and Delbruck 1968).

If a sample is laterally homogeneous, bleaching with an extended periodic pattern is useful, because data can then be gathered simultaneously from many fluorophores, which provides a brighter signal with reduced concomitant photon shot noise.

We derive here an explicit form for periodic sine-wave TIR/FRAP in the presence of reversible nonspecific binding kinetics, which is the situation of interest for the possible reaction rate enhancement of a soluble agonist to specific surface receptors. We confine the derivation to the 'reaction-limited' case, which assumes that $R_R/R_{BND} \rightarrow 0$. Physically, this case means that when a molecule desorbs from the surface, it diffuses away so fast and far from the surface in the timescale of the recovery, that it essentially 'never' comes back. Starting with equation (6.67), we use

$$\zeta(\mathbf{r}) = \frac{1}{2}(1 + \cos px) = \frac{1}{4}[2 + (e^{ipx} + e^{-ipx})] \tag{6.117}$$

for the illumination intensity pattern, with $p = 2\pi/s$, where s is the full wavelength of the pattern, given by equation (5.9) if it was created by interfering coherent beams. We assume that the periodic pattern extends for many pattern wavelengths in the x direction. The integration is straightforward if variable substitutions are made to change the integration variables from $(\mathbf{r},\mathbf{r}')$ to $(\mathbf{u},\mathbf{r}')$:

$$\begin{aligned} \mathbf{u} &\equiv \left(\frac{\mathbf{r}-\mathbf{r}'}{(4D_C t)^{1/2}}\right) = (u_x, u_y) \\ d^2\mathbf{u} &= \left(\frac{d^2\mathbf{r}}{(4D_C t)}\right) \\ \mathbf{r}' &= (x', y') \\ d^2\mathbf{r}' &= dx'dy'. \end{aligned} \tag{6.118}$$

The final result is quite simple:

$$\frac{G(t)}{G(0)} = e^{-R_R t}\frac{2}{3}\left(1 + \frac{1}{2}e^{-p^2 D_C t}\right). \tag{6.119}$$

This result is valid both for TIR/FCS and for shallow-bleach TIR/FRAP. For experiments with extended periodic patterns covering a wide area with very many molecules, TIR/FCS is much less useful than TIR/FRAP because the number fluctuations FCS depends upon are relatively small.

In the language of TIR/FRAP, equation (6.119) shows interesting effects. For $R_R = 0$ (i.e. no exchange between surface and solution), the 'recovery' never achieves 100% because the reservoir of unbleached molecules is not infinite, unlike standard EPI spot FRAP. In fact, it only recovers one third of the bleach depth at $t = 0$. But if $R_R > 0$ and it is fast enough to be seen during the experiment (say, if $R_R > 0.01p^2D_C$) then the recovery approaches 100%, as the infinite reservoir in the bulk solution replaces the bleached fluorophore on the surface. In such cases, it should be easy (by curve fitting or using at least two different periodic spacings) to obtain actual values for both R_R and D_C.

Note that for molecules diffusing in a sinusoidal pattern (equation 6.119), $G(t)$ is a pure exponential decay, but for molecules diffusing in an exponential pattern (considered in section 6.8.7 above), $G(t)$ is a w function.

6.9 TIR-continuous photobleaching

Subsection 6.8.13 describes a TIR/FRAP scheme to measure molecular diffusion coefficients in spherical secretory granules. This scheme is specifically designed for very small spheres, on the order of the wavelength of light, λ_0, and only slightly larger than the evanescent field depth *d*. Many objects of interest in biology (e.g. bacterial cells) are larger (5–10 λ_0) and non-spherical, but still too small for standard EPI spot FRAP to work. If we are interested in bulk diffusion inside a bacterium (rather than diffusion on its surface), a simplifying variation of the mathematical model is available.

Additionally, the variation is coupled to a different experimental protocol for photobleaching that allows the measurement of fast diffusion; the protocol in subsection (b) is somewhat limited in speed because the bleaching pulse needs to be effective in a much shorter time than the characteristic time of diffusion through the bleached region. The protocol considered here involves continuous photobleaching (CP) that takes place simultaneously with observation, rather than a pulse and recovery thereafter. The CP theory has no close analogy with FCS. For this reason, CP is addressed in a separate section outside of section 6.8.

The main modeling difference here is that the photobleaching is approximated to occur only at $z = 0$ in a layer that is infinitesimal compared to the thickness of the cell, rather than a well-defined spread-out imprint based on an exponential decay. Figure 6.28 illustrates this approximation.

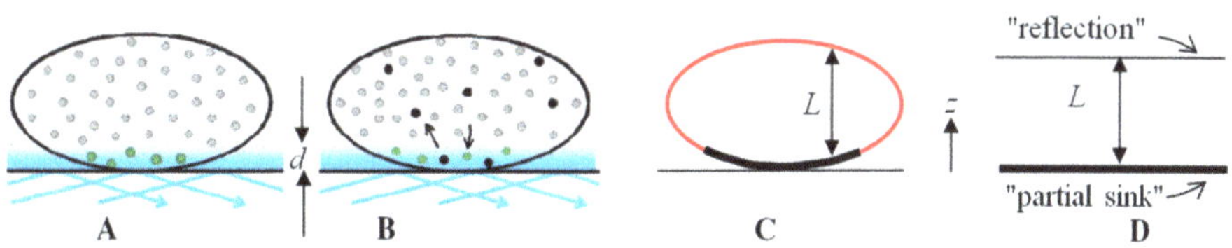

Figure 6.28. Scheme for TIR-CP on a non-spherical (e.g. elliptical, as shown here) cell. The cell is small, perhaps only several λ_0, but has a thickness that is still much larger than the evanescent field depth *d*. Molecules contained within the cell do not fluoresce (gray circles, panel A) until they diffuse into the surface-associated evanescent field. Eventually (panel B), the fluorescent molecules (green circles) are permanently bleached (black circles) by exposure to the evanescent field. The region where bleaching takes place is assumed (panel C) to be only a thin region at the bottom of the cell near the substrate. The thickness of the cell, *L*, varies somewhat over the width of the cell, but is assumed to be a constant near its maximal value as measured independently (say, with confocal microscopy). A further assumption (panel D) flattens the active bleaching region to $z = 0$ and also assumes that the *L* of the cell (and others that may be in the active field of view) is constant everywhere. The diffusion problem is thereby approximated as one dimensional in *z*. The field of view is laterally limited by an image-plane diaphragm, but the field still includes at least a whole single cell and possibly several others nearby. Panels A and B adapted with permission from Slade *et al* (2009), copyright (2009) American Chemical Society.

The illumination is held at a constant intensity during any single experimental run, and the fluorescence it excites is observed continuously. Two competing processes, diffusion and photobleaching, contribute to the rate and shape of fluorescence decay. At low intensities, the decay is determined by the photobleaching rate within the evanescent wave. At high enough intensities, the photobleaching rate is fast enough that the diffusion of unbleached molecules into the evanescent wave becomes the rate-limiting step. By acquiring data in several runs as a function of the excitation intensity, the diffusion coefficient of the fluorescent molecules within the cell can be determined. The corresponding mathematics can be set up as follows.

Let $U(z, t)$ be the concentration of unbleached molecules as a function of (1D) space and time.

The diffusion equation is

$$\frac{\partial}{\partial t}U(z,t) = D_A\frac{\partial^2}{\partial z^2}U(z,t), \tag{6.120}$$

where D_A is the bulk diffusion coefficient inside the cell (following the notation for bulk diffusion already used). The solution is determined by the initial condition and the boundary conditions at $z = 0$ and $z = L$. The initial condition is

$$U(z, 0) = C_A, \tag{6.121}$$

where C_A is the total bulk concentration of fluorescent molecules. The boundary condition at $z = L$ is

$$\left[\frac{\partial}{\partial z}U(z,t)\right]_{z=L} = 0. \tag{6.122}$$

This 'reflection' condition expresses the notion that the flux across the distal boundary of the cell is zero. The boundary condition at $z = 0$ is a 'partial sink':

$$D_A\left[\frac{\partial}{\partial z}U(z,t)\right]_{z=0} = -d\kappa I_0 U(0,t), \tag{6.123}$$

where κ is a proportionality constant describing the photobleaching propensity with units of intensity^{-1} time^{-1}. The flow of U up into solution (toward increasing z) is less than the flow toward the surface because some unbleached molecules are converted to bleached ones there. This equation states that the net flow rate of unbleached molecules is negative (towards $z = 0$). The flow rate is equal to the unbleached-to-bleached conversion rate, which is proportional to the local unbleached concentration, the bleaching propensity, and the total light energy available for bleaching, which is the integral of the evanescent field $I_0\exp(-z/d) = I_0 d$.

The fluorescence observed over time, as excited by the same evanescent field at the same intensity that does the photobleaching, is:

$$F(t) = QI_0 \int_{z=0}^{L} U(z,t)\ e^{-z/d} dz, \tag{6.124}$$

where Q is a proportionality constant. When normalized to the initial fluorescence $F(0)$, the complete solution is a rather complicated-looking infinite series

$$f(t) \equiv \frac{F(t)}{F(0)} = 2cL\sum_{n=1}^{\infty}\frac{(cd + L)x_n^2 - Le^{-L/d}(c^2 + x_n^2)\cos x_n[L]}{(1 - e^{-L/d})(c + c^2 + x_n^2)(L^2 + dx_n^2)x_n^2}\ e^{-D\left(\frac{x_n}{L}\right)^2 t}, \tag{6.125}$$

where the unitless c (proportional to illumination intensity) is:

$$c \equiv \frac{d\kappa I_0 L}{D_A} \tag{6.126}$$

and the unitless x_n are solutions of the equation

$$x_n \tan x_n = c. \tag{6.127}$$

The details of the derivation are to be found in Slade *et al* (2009).

The behavior of equation (6.125) over different illumination intensity ranges is interesting. At the lowest intensities, bleaching is so slow that the bleached and unbleached concentrations have plenty of time to intermix in the cell volume, and therefore the normalized fluorescence just drops exponentially:

$$\text{low intensity:} \quad f(t) = \exp\left(-\frac{\kappa I_0 d}{L} t\right). \tag{6.128}$$

This low-intensity limit contains no dependence on D_A, so it is not useful for measuring D_A.

At higher intensities, D_A does come into play. For intermediate intensities, the first term ($n = 1$) dominates. It contributes the slowest decay component, and if the time resolution of the observation is low enough, only that component is important. In that case,

$$\text{intermediate intensity:} \quad f(t) \approx \exp\left(-D_A\left(\frac{x_1}{L}\right)^2\right). \tag{6.129}$$

Although I_0 does not appear explicitly in equation (6.115), it is still present there because x_1 depends upon c, which in turn depends upon I_0.

At the very highest intensities, every unbleached molecule that enters the evanescent field is bleached before diffusing away (presumably after emitting many detectable fluorescence photons), so that the fluorescence decay depends upon d, D_A, and L, but no longer upon I_0; the effect of I_0 becomes 'saturated out.'

There are several advantages (and some disadvantages) to using TIR-CP rather than TIR/FRAP. (a) Because bright illumination is used constantly, even during observation, the fluorescence intensity is substantial, even for weakly-labeled samples. Even lower-aperture objectives (and consequently, prism-based TIRF) become sufficient. By contrast, TIR/FRAP must reduce its observational intensity to at least one hundredth of the brief bleach-pulse intensity to avoid bleaching during the probe phase, and this often causes a problem with photon shot noise. (b) TIR-CP essentially slows the effect of D_A because the increasing fluorescence 'recovery' it causes kinetically competes against the decreasing fluorescence effect of the continuous bleaching. Therefore, the frame rate of the camera (or bin rate of the single-channel photodetector) need not be especially fast. (c) Because there is a substantial amount of bleaching involved during a TIR-CP experiment, whereby much or most of the fluorophore in the cell eventually becomes bleached, photochemical damage to a living cell may be increased. This possibility has not been investigated. (d) As discussed in section 6.8.13.2, reversible bleaching (the tendency of bleached fluorophore to 'recover' on its own on a fast timescale) may affect the theory of TIR-CP, thus requiring a modification. (e) Interpretation of the results depends upon the thickness of the cell L. Not all areas of the cell may have the same L, so the effects will portray an average that is difficult to define precisely.

6.10 TIR-FRET

Fluorescence resonance energy transfer (FRET) is a technique for visualizing nanometer-scale distances between components of a biological system. It does this by exciting a 'donor' fluorophore on one component and detecting the response of a nearby 'acceptor' on the other component. Aside from localizing the region of interest to the near-substrate zone, the TIR-FRET combination offers some unique features.

The specific binding being measured is often reversible, which means that one or both components might exist at significant concentrations throughout the bulk of the sample. Even if one of the components is occasionally surface bound (either to the substrate or to the 'underside' or edge of an overlying cell), the spectroscopic background of the unbound components may swamp the FRET signal under EPI illumination. TIRF can reduce that background by emphasizing the substrate-proximal component's interactions with its reversibly binding partner.

Another feature of TIR-FRET is actually related to the above subsection on TIR/FRAP. One way of detecting FRET is to flash the illumination at the (short) donor wavelength at which the acceptor is only minimally excited directly, and then seeing whether the acceptor becomes bleached. This happens only if energy is transferred from donor to acceptor. It is helpful that TIR only illuminates a thin layer, because direct bleaching of isolated donor molecules farther out in the solution before they (reversibly) bind to the acceptor would be undesirable; those donors represent a reservoir of viable energy-transfer partners.

A third feature TIR-FRET is the ability to detect macromolecular conformational changes induced by the surface adsorption process itself. This is particularly important in biological processes that are triggered by chemistry at a surface, e.g. blood coagulation.

6.11 Two-photon TIRF

All of the previous sections assume that the probability of fluorophore excitation, and the consequent fluorophore emission power, are proportional to the first power of the local illumination intensity, I. Intuitively, that makes sense: an absorbed photon brings a ground state fluorophore into an excited state; twice the number of available photons leads twice the number of excited fluorophores. The probability of two-photon excitation is proportional to the second power of the intensity, I^2. The efficiency of this two-photon process is very low for dim light, but increases (proportionally to I). It requires a bright light, usually achieved using a short-duration pulse, so that the intensity is high but the short duration limits the deposition of energy (and heat) to the sample.

Schematically, two-photon absorption can be drawn on a standard energy level diagram by introducing an intermediate 'virtual state,' as shown in figure 6.29. This virtual state is not really a single state, but rather a sum over all the states in the system, as derived in the theory section immediately below

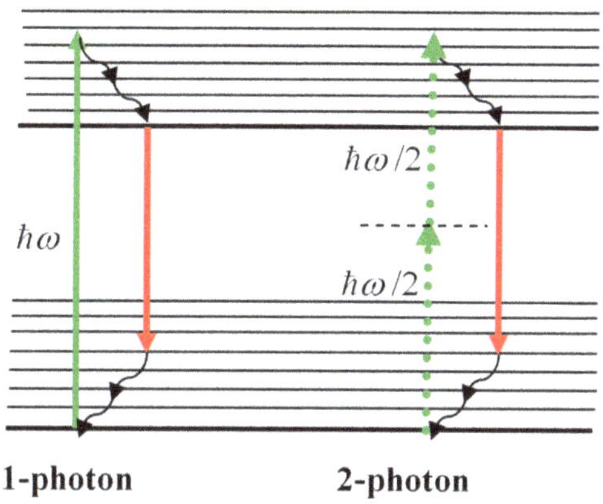

Figure 6.29. Excitation by one photon takes the molecule from near the lowest vibrational level of its ground state to a higher vibrational level of an electronic excited state (solid green arrow). Radiationless transitions (squiggly black lines) return the molecule from higher to lower vibrational levels. Because the difference in the energies of the states involved in the absorption is equaled by the energy of the photon $\hbar\omega$, this mechanism is called 'resonant absorption.' Excitation by two photons (each with half the energy, $\hbar\omega/2$) takes the molecule to an intermediate 'virtual state' (dashed black line) and then, simultaneously, to the same higher state as that achieved through one-photon excitation. The virtual state is shown as a single level for graphical convenience.

6.11.1 Two-photon theory

Two-photon absorption can happen if two photons arrive at the fluorophore close enough to the same time (within 10^{-16} to 10^{-15} s) that their effects—and in particular, their energies—add together. Then, two photons which individually might lack the energy to excite a fluorophore to its lowest excited state act can together to effect an excitation. However, this explanation is obviously very crude. In this section, the outlines of the relevant quantum mechanical theory—time-dependent perturbation theory—are presented. The goal here is not necessarily to calculate the likelihood of such a two-photon event, but rather to show that it can occur at all, and that it has an I^2 dependence. This section is presented for background information. It is not strictly necessary in order to understand the two-photon applications in subsequent sections. The presentation starts from the 'beginning,' (the Schrödinger equation) and follows the introductory level of Griffiths and Schroeter (2018), which should be consulted for more details.

The quantum mechanical approach given here is a 'semiclassical' approximation, in which the energy levels in the matter absorbing the light energy are quantized, but the electric fields themselves are classical and non-quantized. This presentation shows how, to a first-order approximation, 'transition dipole moments' come about, and how, to a second-order approximation, two-photon absorption occurs.

The Schrödinger equation predicts the quantum states that exist in a system, their energies, and the probability of the system actually being observed in a particular state:

$$i\hbar\frac{\partial\Psi}{\partial t} = (H_0 + V)\Psi. \tag{6.130}$$

Here, H_0 is a combination of terms (some second derivatives and some algebraic expressions) that describe the kinetic and potential energy structure of the system (called the 'Hamiltonian operator'), assumed here to be time-independent, and V is a small change—a 'perturbation'—in the energy of the system. V can be a space-and-time-dependent potential energy imposed by the environment (say by a sinusoidal electric field in incident light); the form of Ψ can be a function of space and time, but, in general, could also involve the spin states of the system, which we do not pursue here.

If $V = 0$, Ψ is any single state (typically one of many discrete states Ψ_n) that solves equation (6.130) with $V = 0$. Those states have constant energies E_n, according to the equation

$$H_0\Psi_n = E_n\Psi_n \tag{6.131}$$

and each Ψ_n thereby satisfies the equation

$$i\hbar\frac{\partial\Psi_n}{\partial t} = E_n\Psi_n$$

so that

$$\Psi_n = \hat{\Psi}_n e^{-iE_nt/\hbar} \tag{6.132}$$

in which $\hat{\Psi}_n$ (with its superscript 'hat') has only a space dependence, but not a time dependence.

With V 'turned on,' Ψ is altered so as to be a sum over all the non-perturbed states:

$$\Psi = \sum_n c_n(t)\Psi_n = \sum_n c_n(t)\hat{\Psi}_n e^{-iE_nt/\hbar}. \tag{6.133}$$

The coefficients c_n can be complex, and can be functions of time if V is a function of time. $|c_n|^2$ is the (time-varying) probability that the system will be found in state Ψ_n at any particular time t. Finding an expression for $|c_n|^2$ is the goal of this derivation.

Plugging equation (6.133) into (6.130) gives

$$i\hbar\sum_n \hat{\Psi}_n \frac{\partial c_n(t)e^{-iE_nt/\hbar}}{\partial t} = \sum_n c_n(t)(H_0 + V)\hat{\Psi}_n e^{-iE_nt/\hbar}. \tag{6.134}$$

The left-hand side of equation (6.134) gives:

$$\begin{aligned} i\hbar\sum_n \hat{\Psi}_n e^{-iE_nt/\hbar}\left[\frac{\partial c_n(t)}{\partial t} - c_n(t)\frac{iE_n}{\hbar}\right] &= i\hbar\sum_n \hat{\Psi}_n e^{-iE_nt/\hbar} \\ &\quad \frac{\partial c_n(t)}{\partial t} + \sum_n c_n(t)\hat{\Psi}_n e^{-iE_nt/\hbar}E_n. \end{aligned} \tag{6.135}$$

The right-hand side of equation (6.134) gives:

$$\sum_n c_n(t)(H_0 + V)\hat{\Psi}_n e^{-iE_n t/\hbar} = \sum_n c_n(t)\hat{\Psi}_n e^{-iE_n t/\hbar} E_n + V\sum_n c_n(t)\hat{\Psi}_n e^{-iE_n t/\hbar}. \quad (6.136)$$

After the cancellation of identical terms on the left and right sides, what remains is:

$$i\hbar\sum_n \hat{\Psi}_n e^{-iE_n t/\hbar}\frac{\partial c_n(t)}{\partial t} = V\sum_n c_n(t)\hat{\Psi}_n e^{-iE_n t/\hbar}. \quad (6.137)$$

To separate out an equation for each of the different c_n factors, we make use of the fact that any two solutions of equation (6.131) are 'orthonormal':

$$\oint \hat{\Psi}^*_j \hat{\Psi}_n d^3\mathbf{r} = \delta_{jn}, \quad (6.138)$$

where δ_{jn} is the Kronecker delta (=1 for $j = n$, 0 otherwise), and the integration is over all space and spin variables. Multiplying equation (6.137) by any particular $\hat{\Psi}^*_j$ and integrating then gives:

$$\frac{\partial c_j(t)}{\partial t} = -\frac{i}{\hbar}\sum_n c_n(t)\, e^{-i(E_j - E_n)t/\hbar} \oint \hat{\Psi}^*_j V \hat{\Psi}_n d^3\mathbf{r}. \quad (6.139)$$

The perturbation V essentially connects the states $\hat{\Psi}_j$ and $\hat{\Psi}_n$ so that an integral involving their product is no longer zero. Note that the summation over n includes *all* the states, including j.

The integral involving V takes a special form (and special properties) if the perturbation originates from an electric field **E** acting on an electron (any of those in the orbitals of the molecule). The potential energy imposed by **E** is just the force multiplied by the position **r** through which the electron moves. Therefore,

$$V = q_e e^{i\omega t}\mathbf{E}\cdot\mathbf{r}, \quad (6.140)$$

where q_e is the electron charge and ω is the angular frequency of the incident light. When installed into the integral of equation (6.139), the only factor remaining in the integral (taken over space) is the **r** vector; the other factors do not depend on space, so they just multiply the integral:

$$\oint \hat{\Psi}^*_j V \hat{\Psi}_n d^3\mathbf{r} = q_e e^{i\omega t}\mathbf{E}\cdot\mathbf{M}_{jn} \quad (6.141)$$

where

$$\mathbf{M}_{jn} \equiv \oint \hat{\Psi}^*_j \mathbf{r} \hat{\Psi}_n d^3\mathbf{r}. \quad (6.142)$$

$\mathbf{M}_{jn}$ is called the 'transition dipole moment' vector. Note that if $j = n$, then $\mathbf{M}_{jn} = 0$ because the integrand of equation (6.142) becomes a product of an even function (in space) $|\hat{\Psi}_n|^2$ and the odd function **r**.

As a system of potentially many coupled equations, equation (6.139) is hard to solve. Instead, we use equation (6.139) as a guide to setting up a formula for successive approximations. Rewriting it as such, and in the particular form of V involving the transition dipole moment, we get

$$\frac{\partial c_j^{(k+1)}(t)}{\partial t} = -\frac{iq_e}{\hbar}\sum_n c_n^{(k)}(t)\, e^{-i(E_j-E_n-\hbar\omega)t/\hbar}\mathbf{E}\cdot\mathbf{M}_{jn}. \tag{6.143}$$

Here, equation (6.139) has become a recursive relationship. The k superscripts in parentheses refer to the order of approximation. Given kth-order approximations for all the c_n on the right, equation (6.143) prescribes how to obtain the next higher-order approximation for a particular c_j on the left. This is where the first-order and second-order approximations (corresponding to one- and two-photon absorption) enter the picture. To start off this successive approximation process, we assume that the system is in a very particular definite state a (say, the ground state) just before the perturbation V is turned on. Then, to the zeroth order (denoted by the superscript in parentheses), we assume

$$\begin{aligned} c_a^{(0)}(t) &= 1 \quad \text{and} \\ c_n^{(0)}(t) &= 0 \quad \text{for all } n \neq a. \end{aligned} \tag{6.144}$$

This is clearly a crude approximation, because we do not expect these values to be true for all time t, only for $t = 0$. However, we proceed anyway to insert the equation (6.144) values into the right-hand side of equation (6.143) to see how that equation further refines them. Substituting the particular subscript a for the general subscript j, and noting that $W_{aa} = 0$, we obtain:

$$\frac{\partial c_a^{(1)}(t)}{\partial t} = 0 \quad \text{and therefore} \quad c_a^{(1)}(t) = 1. \tag{6.145}$$

Substituting the particular subscript b ($\neq a$) (say, for a higher excited state) for the general subscript j, we obtain

$$\frac{\partial c_b^{(1)}(t)}{\partial t} = -\frac{iq_e}{\hbar}\mathbf{E}\cdot\mathbf{M}_{ba}e^{-i(E_b-E_a-\hbar\omega)\,t/\hbar}, \tag{6.146}$$

and integrating over time (and noting that $\mathbf{E}\cdot\mathbf{M}_{ba}$ is time-independent) gives

$$c_b^{(1)}(t) = -\frac{iq_e}{\hbar}\mathbf{E}\cdot\mathbf{M}_{ba}\int_0^t e^{-i(E_b-E_a-\hbar\omega)t'/\hbar}dt'. \tag{6.147}$$

We are especially interested in $c_b^{(1)}(t)$ because its square is proportional to the first-order probability that a system that started in state a is later observed to be in state b, to the first order:

$$|\,c_b^{(1)}(t)\,|^2 = \frac{q_e^2}{\hbar^2}\,|\,\mathbf{E}\cdot\mathbf{M}_{ba}\,|^2\left|\int_0^t e^{-i(E_b-E_a-\hbar\omega)t'/\hbar}dt'\right|^2. \tag{6.148}$$

Equation (6.148) exhibits a number of important features of single-photon excitation:

(i) The probability of first-order absorption from state a to b is proportional to $\cos^2\theta$, where θ is the angle between the electric field polarization of the light and the transition dipole moment. (We used this fact frequently in all the previous discussions of polarization in TIR excitation.)

(ii) The probability of first-order transition is maximal when the incident photon energy $\hbar\omega$ equals the difference between the excited-state energy E_b and the initial-state energy E_a ('resonance').

(iii) When the illumination frequency meets that resonance condition, the integral-squared term in equation (6.134) becomes proportional to the time t for small values of the exponent; i.e. the (presumably small) probability of excitation increases linearly with the duration of illumination.

(iv) The transition dipole moment $\mathbf{M}_{jn}$ is a geometrical property of the two states $\hat{\Psi}^*{}_j$and$\hat{\Psi}_n$, presumably fixed to the molecule even if it rotates. If both states have the same inversion geometrical symmetry ($\mathbf{r} \rightarrow -\mathbf{r}$)), both either odd or even, then their product is even, the $\mathbf{r}$ factor in the integrand makes the whole integrand odd, and then $\mathbf{M}_{jn} = 0$. Such transitions are thereby 'forbidden.'

(v) Most important for the present discussion, the probability of absorption is proportional to $|\mathbf{E}|^2$; i.e. proportional to the *first* power of intensity I.

To obtain the second-order transition probability for any state b, we rewrite equation (6.147) in terms of a general subscript n:

$$c_n^{(1)}(t) = -\frac{iq_e}{\hbar}\mathbf{E} \cdot \mathbf{M}_{na}\int_0^t e^{-i(E_n-E_a-\hbar\omega)t'/\hbar}dt' \tag{6.149}$$

and insert that into the right-hand-side summation of the recursion relationship (equation 6.143):

$$\begin{aligned} \frac{\partial c_b^{(2)}(t)}{\partial t} &= -\frac{iq_e}{\hbar}\sum_n c_n^{(1)}(t)\, e^{-i(E_b-E_n-\hbar\omega)t/\hbar}\mathbf{E} \cdot \mathbf{M}_{bn} \\ \frac{\partial c_b^{(2)}(t)}{\partial t} &= -\frac{q_e^2}{\hbar^2}\sum_n(\mathbf{E} \cdot \mathbf{M}_{na})(\mathbf{E} \cdot \mathbf{M}_{bn})e^{-i(E_b-E_n-\hbar\omega)t/\hbar} \\ &\quad \int_0^t e^{-i(E_n-E_a-\hbar\omega)t'/\hbar}dt'. \end{aligned} \tag{6.150}$$

Integrating this over time and then squaring it gives the following result:

$$\begin{aligned} c_b^{(2)}(t) &= -\frac{q_e^2}{\hbar^2}\sum_n(\mathbf{E} \cdot \mathbf{M}_{na})(\mathbf{E} \cdot \mathbf{M}_{bn})\int_0^t e^{-i(E_b-E_n-\hbar\omega)t''/\hbar}\int_0^{t''} e^{-i(E_n-E_a-\hbar\omega)t'/\hbar}dt'dt'' \\ |\, c_b^{(2)}(t)\,|^2 &= \frac{q_e^4}{\hbar^4}\left|\sum_n(\mathbf{E} \cdot \mathbf{M}_{na})(\mathbf{E} \cdot \mathbf{M}_{bn})\int_0^t e^{-i(E_b-E_n-\hbar\omega)t''/\hbar}\int_0^{t''} e^{-i(E_n-E_a-\hbar\omega)t'/\hbar}dt'dt''\right|^2. \end{aligned} \tag{6.151}$$

This last expression, describing the second-order correction to the probability of excitation by a sinusoidal **E** field, looks rather complicated. However, it has the features we are seeking that allow two-photon excitation:

- The electric field appears to the fourth power, $|\mathbf{E}|^4$, which corresponds to the second power of intensity, I^2.
- Starting at state a, the system uses all its states n to 'sense' the incident field on its course to making a landing in state b. This indirect route, with multiple states n as intermediates, is often pictured as a quick stop at a single 'virtual' state in between the real states a and b. But apart from convenience in description, a single virtual state does not really exist.
- The polarization effect is complicated. Direct $a \rightarrow b$ single-photon transitions that cannot occur because of orthogonal directionality ($\mathbf{E} \cdot \mathbf{M}_{ba} = 0$) are still possible in two-photon excitations because of the $\mathbf{M}_{na}$ and $\mathbf{M}_{bn}$ intermediates. Likewise, single-photon transitions that are quantum mechanically forbidden by the symmetry ($\mathbf{M}_{ab} = 0$) are still allowable in the two-photon world.
- The time dependence in equation (6.141) can be explicitly integrated and shows that two-photon absorption probability is maximized when $E_b - E_a = 2\omega$, which is why an incident frequency corresponding to half the energy gap between states a and b is used in two-photon microscopy.

6.11.2 Reduction of scattering effect

Given that two-photon excitation is proportional to I^2, the two-photon efficiency for an evanescent field is proportional to:

$$(I_0 e^{-z/d})^2 = I_0^2 e^{-2z/d} = I_0^2 e^{-z/(d/2)}, \tag{6.152}$$

so the characteristic depth d is halved. However, there are restrictions on the incident wavelengths that are usable. If the same wavelength (or a shorter wavelength) is used that could be used for one-photon excitation, then one-photon excitation and fluorescence still occur simultaneously, swamping the two-photon effect. Also, living cells are most susceptible to photodamage at short wavelengths, and increasing the intensity enough to achieve two-photon effects would be detrimental to survival. Therefore, long-wavelength IR light is used, with an energy-per-photon ($h\nu = 2\pi h/\lambda$) that is insufficient to cause one-photon excitation. Its wavelength is generally around twice that of the usual visible one-photon excitation standard in fluorescence microscopy. Thus the advantage of halving d (based on the I^2 dependence) is completely eliminated by the doubling of d due to the longer wavelength.

The real advantage of two-photon TIRF is its relative reduction in the effect of scattering. Using the rough approximation (equation 3.59) that the scattering is a constant term added to the evanescent exponential, squaring the total intensity gives

$$\begin{aligned} &[(1 - \alpha_{\text{scatt}})e^{-z/d} + \alpha_{\text{scatt}}]^2 \\ &= (1 - \alpha_{\text{scatt}})^2 e^{-2z/d} + 2\alpha_{\text{scatt}}(1 - \alpha_{\text{scatt}})e^{-z/d} + \alpha_{\text{scatt}}^2. \end{aligned} \tag{6.153}$$

For α_{scatt} = 0.1 (as suggested after equation (3.60)), one-photon excitation allows scattering to contribute ~11% of the fluorescence intensity at $z = 0$. Of course, this percentage is higher at larger z, which represents the compromised surface selectivity. But two-photon excitation has three effects corresponding to the three terms in equation (6.153):

(i) A fast-decaying exponential, which decays at twice the rate of one-photon excitation. This term accounts for 81% of the total at $z = 0$.

(ii) A slower-decaying exponential, equal to the decay rate of one-photon excitation. This term contributes ~18% of the total at $z = 0$. Although this slower-decaying $e^{-z/d}$ contribution is extends farther out in z than the scattering-free term $e^{-2z/d}$, at least it does decay, so surface selectivity is not seriously compromised.

(iii) A constant term that contributes only 1% of the intensity at $z = 0$. However, this z-independent constant term, i.e. the term that could seriously compromise surface selectivity, is much reduced compared to the one-photon case.

Another advantage of two-photon over one-photon excitation appears in the configurations in which the illumination region is laterally limited by focusing or interference (see section 5.11), and the region's edges are 'fuzzy.' Using the same mechanism (as above) that suppresses the impact of more dimly illuminated regions, two-photon TIRF produces a more confined lateral extent.

6.11.3 Requirement for high intensity

In general, two-photon efficiency needs high-intensity IR illumination. This can be achieved best by lateral region-confining configurations. However, to achieve TIR, those same configurations can involve blocking much of the incident light that was heading along subcritical paths, so that the necessary blocking partially counters the advantage of small lateral confinement.

Another possible route to higher intensities is to employ a metal-film coated surface and inject the illumination beam at the surface plasmon angle (see section 3.3 and also figure 2.2). This can increase the illumination intensity by a factor of three to four. If the emitted fluorescence excites a surface plasmon effect in the metal film, the collection efficiency of the emission can also increase. However, a metal film quenches emission near $z = 0$ (within the first nanometer or so), so surface plasmons are best utilized for fluorophores outside that range (He *et al* 2009).

A standard technique in two-photon fluorescence excitation is to shorten the duration of illumination using a fast-pulse laser. This allows a brighter intensity during the pulse, while still limiting the total average power to the sample, to minimize heating and photodamage. However, a short (femtosecond) pulse necessarily contains a spread of frequencies. Because of all the optical media and surfaces in the path, chromatic dispersion broadens the pulse and thereby decreases peak intensity: the pulses of some colors make it through the system more quickly than others. To counteract this undesirable effect, pulse-shaping techniques have been employed in conjunction with two-photon TIRF (Lane *et al* 2012).

A two-photon system enjoys a large separation of the IR illumination's long wavelength from the visible fluorescence emission's short wavelength. This allows the optical filtering to be more capable of capturing all of the fluorescence spectral band.

6.11.4 Two-photon sine-wave-pattern TIRF

Two-photon TIRF excitation can be applied to sine-wave-pattern FRAP, which was discussed in its single-photon case in section 6.8.17. In principle, the pulse train that produces two-photon effects can be deployed either during the 'bright' bleach phase, or during the 'dim' probe phase, or both. (The 'both' case is mainly discussed here.) In practice, 'bright' and 'dim' should be defined as the number of rapidly successive IR pulses in a given time duration, during which the bleach phase receives at least a hundred times as many pulses as the probe phase per sample bin time. The two-photon aspect introduces some new features.

One such feature is a change in the shape of the postbleach fluorescence recovery. The sine-wave intensity pattern $\zeta(\mathbf{r})$ is assumed to be produced by intersecting coherent (monochromatic) beams (although monochromaticity is somewhat reduced for very short pulses)

$$\zeta(\mathbf{r}) = \frac{1}{2}(1 + \cos px), \tag{6.154}$$

so that the effective two-photon intensity is

$$\zeta^2(\mathbf{r}) = \frac{1}{4}(1 + \cos px)^2 = = \frac{1}{8}(3 + 4\cos px + \cos 2px). \tag{6.155}$$

Squaring the intensity produces a spatial frequency ($2p$) component contribution that is double that of the single-photon frequency component (p) intensity. Since the shape of the single-photon recovery follows the form (as derived in section 6.8.15)

$$\frac{G(t)}{G(0)} \sim \left(1 + \frac{1}{2}e^{-p^2 D_c t}\right), \tag{6.156}$$

the new $2p$ contribution introduces a component of the recovery at four times the rate ($4p^2$). The faster decay allows slower diffusion rates to be detectable in the same experimental timescale.

Another feature relates to the size of the 'reservoir' from which unbleached molecules diffuse into probe-illuminated regions, thereby providing a fluorescence 'signal.' Unlike spot FRAP, the reservoir is not infinite in sine-pattern FRAP, and in fact, the fluorescence recovery can restore, at best, only one third of the amount lost during the bleach phase (as previously shown). A deeper bleach can increase the 'amount lost' and thereby provide more absolute signal. However, a deeper bleach also reduces the one-third fraction, because it reduces the reservoir of unbleached molecules. It turns out that the absolute signal reaches a maximum at a moderate bleach depth, namely at 22% of the prebleach fluorescence. However, two-photon illumination has narrower periodic peaks (see figure 6.30), so the reservoir of

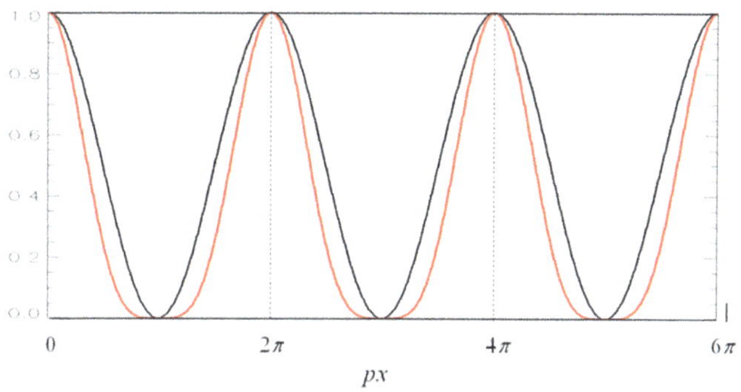

Figure 6.30. Comparison of ζ (black) vs ζ^2 (red) for a cosine wave, equations (6.154) and (6.155).

unbleached fluorophore is broader. The absolute recovery amplitude can then be as large as 36% of the prebleach fluorescence, which improves the signal/noise ratio.

The last feature concerns binding kinetics. One mode for recovery is adsorption/desorption from the TIR surface. If that can occur, then the reservoir again becomes infinite. In one-photon-pattern TIRF experiments, the adsorption/desorption kinetic rate influences the recovery more because it dominates the relatively small lateral (surface diffusion) recovery. But in two-photon pattern TIRF, lateral diffusion has a stronger relative influence.

Because of the sharpening of the sine-wave pattern by the two-photon effect, sine-wave-based structured illumination can be pushed to yield an even higher lateral resolution enhancement than it would in the single-photon case. Although described for non-TIR optics (Gustafsson 2005, Ingerman *et al* 2019), the approach should also work for the TIR case. In addition, the two-photon TIR optics would suppress the contribution from scattering.

6.11.5 Two-photon excitation with slab waveguides

As shown in section 3.4, a waveguide (in that case, a slab) can support very strong 'resonances,' where the multiple reflections add up with the correct phases to produce large fields inside the waveguide and consequently large evanescent fields at their surface. This enhancement is strong enough to produce two-photon effects. In the case of resonance in planar slabs, this uniquely produces two-photon excitation over a large area, not just in a focused beam, which is useful for non-microscopic biosensors. In addition, the near field of the emission from fluorophores on all points on the surface is trapped by multiple total reflections in the slab, ultimately propagating toward an end of the slab and observed there at high efficiency by a photodetector.

The creation of a very bright evanescent field by resonance in a slab can also be deployed in a microscopic configuration, where the surface of the slab is essentially the coverslip upon which the molecular or cellular sample is adhered, and two-photon excited TIRF can be observed over a wide area without scanning from above or below through a microscope objective. The result is wide-field illumination, selective surface illumination, and background suppression, all in one step.

Further reading

Abbineni P S, Briguglio J S, Chapman E R, Holz R W and Axelrod D 2021 Vamp2 and synaptotagmin mobility in chromaffin granule membranes: implications for regulated exocytosis *Mol. Biol. Cell* **1** mbcE21100494

Abbineni P S, Briguglio J S, Chapman E R, Holz R W and Axelrod D 2022 Vamp2 and synaptotagmin mobility in chromaffin granule membranes: implications for regulated exocytosis *Mol. Biol. Cell* **33** ar53

Abramowitz M and Stegun I (ed) 1972 *Handbook of Mathematical Functions* (Washington, DC: National Bureau of Standards)

Adam G and Delbruck M 1968 Reduction of dimensionality in biological diffusion processes *Structural Chemistry and Molecular Biology* ed A Rich and N Davidson (New York: W.H. Freeman & Co.) pp 198–215

Allersma M W, Bittner M A, Axelrod D and Holz R W 2006 Motion matters: secretory granule motion adjacent to the plasma membrane and exocytosis *Mol. Biol. Cell* **17** 2424–38

Anantharam A, Bittner M A, Aikman R L, Stuenkel E L, Schmid S L, Axelrod D and Holz R W 2011 A new role for the dynamin GTPase as a timer that controls fusion pore expansion *Molec. Biol. Cell* **22** 1–13

Anantharam A, Onoa B, Edwards R H, Holz R W and Axelrod D 2010 Localized plasma membrane topological changes upon exocytosis visualized by polarized-TIRFM *J. Cell Biol.* **188** 415–28

Axelrod D 1979 Carbocyanine dye orientation in red cell membrane studied by microscopic fluorescence polarization *Biophys. J.* **26** 557–74

Axelrod D and Omann G M 2006 Combinatorial microscopy *Nat. Rev. Mol. Cell Biol* **7** 944–52

Bekshaev A Y, Bliokh K Y and Nori F 2013 Mie scattering and optical forces from evanescent fields: a complex-angle approach *Opt. Express* **21** 7082–95

Bohannon K P, Holz R W and Axelrod D 2017 Refractive index imaging of cells with variable-angle near TIR microscopy *Microsc. Microanal.* **23** 978–88

Duveneck G L, Bopp M A, Ehrat M, Balet L P, Haiml M, Keller U, Marowsky G and Soria S 2003 Two-photon fluorescence excitation of macroscopic areas on planar waveguides *Biosens. Bioelectron.* **18** 503–10

Fulbright R M and Axelrod D 1993 Dynamics of nonspecific adsorption of insulin to erythrocyte membranes *J. Fluor.* **3** 1–16

Goodman J W 1968 *Introduction to Fourier Optics* (New York: McGraw-Hill)

Griffiths D J and Schroeter D F 2018 *Introduction to Quantum Mechanics* 3rd edn (Cambridge: Cambridge University Press)

Gustafsson M G L 2005 Nonlinear structured-illumination microscopy: wide-field fluorescence imaging with theoretically unlimited resolution *Proc. Natl. Acad. Sci.* **102** 13081–6

He R-Y, Su Y-D, Cho K-C, Lin C-Y, Chang N-S, Chang C-H and Chen S-J 2009 Surface plasmon-enhanced two-photon fluorescence microscopy for live cell membrane imaging *Opt. Express* **17** 5987–97

Huang Z and Thompson N L 1993 Theory for two-photon excitation in pattern photobleaching with evanescent illumination *Biophys. Chem.* **47** 241–9

Ingerman E A, London R A, Heintzmann R and Gustafsson M G L 2019 Signal, noise and resolution in linear and nonlinear structured-illumination microscopy *J. Microsc.* **273** 3–25

Katchalski T, Soria S, Teitelbaum E, Friesem A A and Marowsky G 2004 Two photon fluorescence sensors based on resonant grating waveguide structures *Sens. Actuators* B **107** 121–5

Lakowicz J R, Gryczynski Z and Gryczynski I 1996 On the possibility of evanescent wave excitation distal from a solid-liquid interface using light quenching *Photochem. Photobiol.* **64** 636–41

Lane R S K, Macpherson A N and Magennis S W 2012 Signal enhancement in multiphoton TIRF microscopy by shaping of broadband femtosecond pulses *Opt. Express* **20** 25948–59

Ngatchou-Weiss A, Bittner M A, Holz R W and Axelrod D 2014 Protein mobility within secretory granules *Biophys. J.* **107** 16–25

Oheim M and Schapper F 2005 Non-linear evanescent-field imaging *J. Phys. D: Appl. Phys.* **38** R185–97

Slade K M, Steele B L, Pielak G J and Thompson N L 2009 Quantifying green fluorescent protein diffusion in *Escherichia coli* by using continuous photobleaching with evanescent illumination *J. Phys. Chem.* B **113** 4837–45

Starr T E and Thompson N L 2001 Total internal reflection with fluorescence correlation spectroscopy: combined surface reaction and solution diffusion *Biophys. J.* **80** 1575–84

Sund S E, Swanson J A and Axelrod D 1999 Cell membrane orientation visualized by polarized total internal reflection fluorescence *Biophys. J.* **77** 2266–83

Thompson N L, Burghardt T P and Axelrod D 1981 Measuring surface dynamics of biomolecules by total internal reflection with photobleaching recovery or correlation spectroscopy *Biophys. J.* **33** 435–54

Todd I, Mellor J S and Gingell D 1988 Mapping cell-glass contacts of *Dictyostelium* amoebae by total internal reflection aqueous fluorescence overcomes a basic ambiguity of interference reflection microscopy *J. Cell Sci.* **89** 107–14

www.ingramcontent.com/pod-product-compliance
Ingram Content Group UK Ltd.
Pitfield, Milton Keynes, MK11 3LW, UK
UKHW051853100926
8556IPUK00005B/603

9 780750 333498